Variable Air Volume Systems

By

LEO A. MEYER

H. LYNN WRAY, P.E., Technical Advisor

INDOOR ENVIRONMENT TECHNICIANS LIBRARY

FOREWORD

You are probably working as a technician in one of the indoor environment fields. However, don't fall into the trap of thinking, "I know all this stuff."

Read each chapter. Then do the Review at the end of the chapter. In my experience, every time I studied material I "knew all about," I learned new ideas and corrected misunderstandings.

If you study each chapter carefully, you will gain new ideas. More important, you will give yourself a solid understanding of different types of VAV systems that you will find in the field.

Indoor Environment Technician's Library

This book is part of the *Indoor Environment Technician's Library.* These are practical books that you can use for training or as reference. These books apply to all areas of the indoor environment industry:

Heating, ventilating, and air conditioning
Energy management
Indoor air quality
Service work
Testing, adjusting and balancing

If You are Training Others

If you are a supervisor training others, you will find that the *Indoor Environment Technician's Library* can make it easier. A Supervisor's Guide is available for each book. It includes teaching suggestions and key questions you can ask to make sure the student understands the material.

Leo A. Meyer

ISBN 0-88069-021-6

TABLE OF CONTENTS

LAMA Books
Leo A. Meyer Associates Inc.
20956 Corsair Blvd
Hayward CA 94545-1002
510-785-1091
510-785-1099 Fax
888-452-6244
lama@best.com

1 VAV SYSTEMS

This book assumes that you know the basics of how HVAC (heating, ventilating, air conditioning) systems work. It describes only VAV (variable air volume) systems. This book contains nine chapters:

- Chapter 1 is background information and general information that applies to all VAV systems.
- Chapters 2 to 8 each deal with a basic type of VAV system. Each of these chapters is organized in the same way:

  A description of the system.

  A schematic drawing of the terminal unit and a description of its operation.

  A schematic drawing of the central air handling unit and a description of its operation.

  A discussion of typical advantages and disadvantages for the particular type of VAV system.

- Chapter 9 describes methods of fan control for VAV system fans.

You must understand the symbols, terms, and abbreviations used in this book. All of them are identified in the text or on the drawings. If you need to review the meaning of an item, **use the Appendix in the back of this book**. It includes a glossary of terms used in VAV systems, and a list of

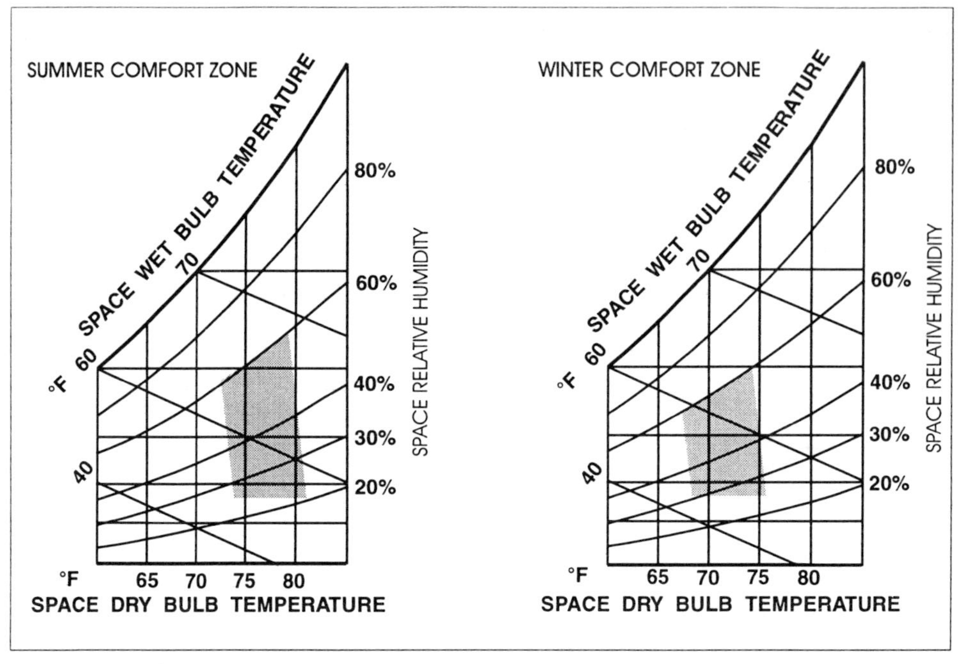

Fig. 1: Comfort zones (ASHRAE 55)

symbols and abbreviations used in the text and drawings for this book. Be sure to refer to the Appendix if you are not sure of the meaning of some term.

THE HVAC SYSTEM

The main goal of any HVAC system is to provide comfort. ASHRAE (American Society of Heating, Refrigerating, and Air-Conditioning Engineers) has established room conditions that are acceptable for most building occupants. Comfort in a space depends on a combination of **temperature** and **humidity** (a measure of the moisture content in the air). Figure 1 shows the comfort zones for summer and winter as defined in ASHRAE Standard 55. The comfort zone is the range of conditions that satisfy most people who are appropriately dressed and are doing light work, such as office work. People doing heavy work and those wearing heavier clothing may need cooler conditions.

The summer and winter comfort zones are slightly different because people dress differently for each season.

Ventilation (outside air) must be provided to every occupied building. Supplying a certain amount of outside air prevents the indoor air from becoming stale and unhealthful. Building codes specify the amount of ventilation air that must be introduced by the HVAC system.

An HVAC system conditions the supply air to provide the spaces with an acceptable combination of humidity and temperature within the comfort zone. It also provides a sufficient amount of outside air for ventilation. The effectiveness of the system depends upon two factors:

❏ The quantity of air being supplied, measured in CFM (cubic feet per minute)

❏ The temperature of the supply air

To heat or cool a space, these two factors are combined in different ways depending upon the type and design of the particular HVAC system. The combinations are:

❏ Constant Volume—Variable Temperature (CV-VT)

❏ Variable Volume—Constant Temperature (VV-CT)

❏ Variable Volume—Variable Temperature (VV-VT)

As the indoor air temperature varies, the **humidity** also increases or decreases. The indoor humidity generally remains within the comfort zone. However in dry climates, many HVAC systems have a humidifier unit in the central air handler to increase the moisture level in the conditioned air when it is needed. In humid climates, systems may have a means of removing moisture from the supply air.

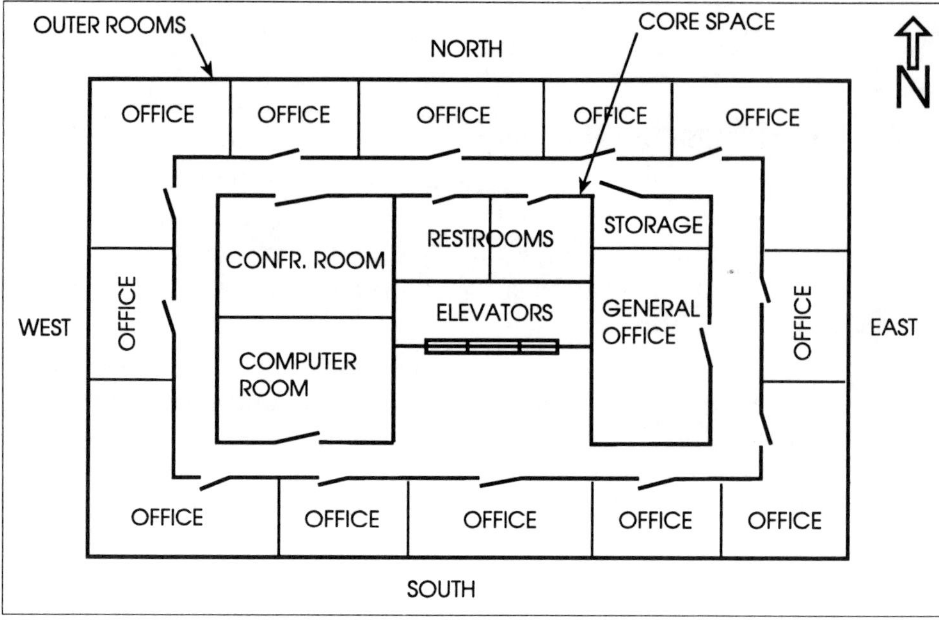

Fig. 2: Zones of a typical office building

DIFFERING ZONES

In large buildings, the HVAC system must meet the varying needs of different spaces. Different zones of a building have different heating and cooling needs.

In the past, large commercial buildings were usually designed with a central light well. This admitted light and air to the inner rooms. As building and land costs increased and HVAC systems developed, this inner light well was eliminated. Now buildings are designed with a **core** of inner rooms where the light well used to be (Fig. 2). The development of this inner core created a new air conditioning problem. The core spaces do not require heating, but do require cooling and ventilation.

The rooms along the outer walls require either heating or cooling as well as ventilation. These **outer rooms** are also called **perimeter spaces**.

Consider the needs of different zones in a typical office building which have different **heat loss** and **heat gain** characteristics. The offices are arranged around the outer walls of the building, with an inner core of rooms that have no outside exposure (Fig. 2).

The **outer rooms** have at least one wall exposed to the outside temperatures. This means that these offices have more **heat loss** and **heat gain** than the interior rooms:

- On cold days, heat transfers from the heated spaces of the building to the colder outside air (heat loss).
- On warm days, heat transfers from the warm outside air to the cooler spaces of the building (heat gain).

The **outer rooms** gain or lose heat at a varying rate. For example, when the sun strikes one side of a building, that side has more heat gain than the sides that are shaded. The position of the sun, color of the wall, insulation, amount of glass, and shading all affect this solar heat gain.

The **core of the building** gains heat from the interior load such as people, lights, and equipment. Therefore, these spaces generally do not require heating except for a top floor where heat is lost through the roof. The normal condition is that they gain too much heat and therefore require cooling when occupied.

The cooling load for both core spaces and outer spaces depends on many factors such as these:

- Types of occupancy (density, active or passive work tasks)
- Heat produced by equipment
- Type and level of lighting

Since different spaces have different rates of heat gain and heat loss, it is impossible for an HVAC system that delivers the same temperature of air at a fixed volume to every space to provide comfort conditions for all of them. Therefore,

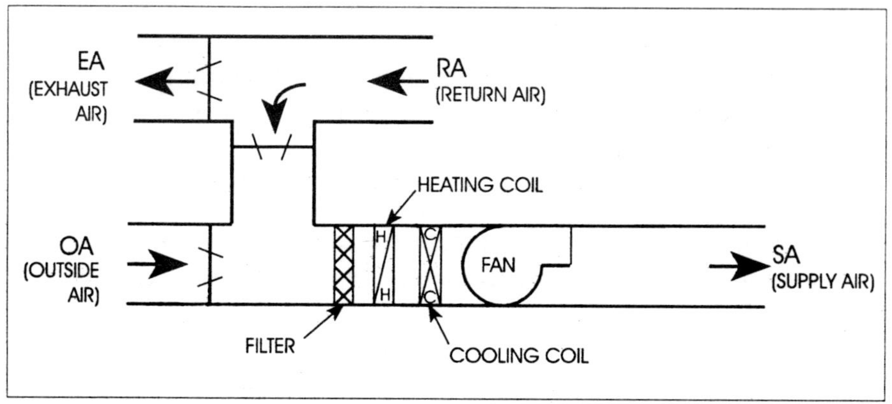

Fig. 3: A single zone system

heating and cooling must be supplied at varying rates to different zones of the building. A **zone** is a space or group of spaces in a building with similar requirements for heating and cooling. All rooms in a zone can be supplied with the same temperature supply air at the same flow rate.

SINGLE ZONE SYSTEM

The first HVAC systems were **single zone systems** (Fig. 3). These systems treat the building as a single zone. They deliver air from a central air handling unit at a constant volume and a varying temperature (CV-VT). The **central air handling unit** in any HVAC system is the equipment in the mechanical room or on the roof that conditions the primary supply air and delivers it to the spaces.

The single zone system is still used in residences and in small commercial buildings. The single zone system is effective in a small building because the heat gain and heat loss for different areas may not vary a great deal.

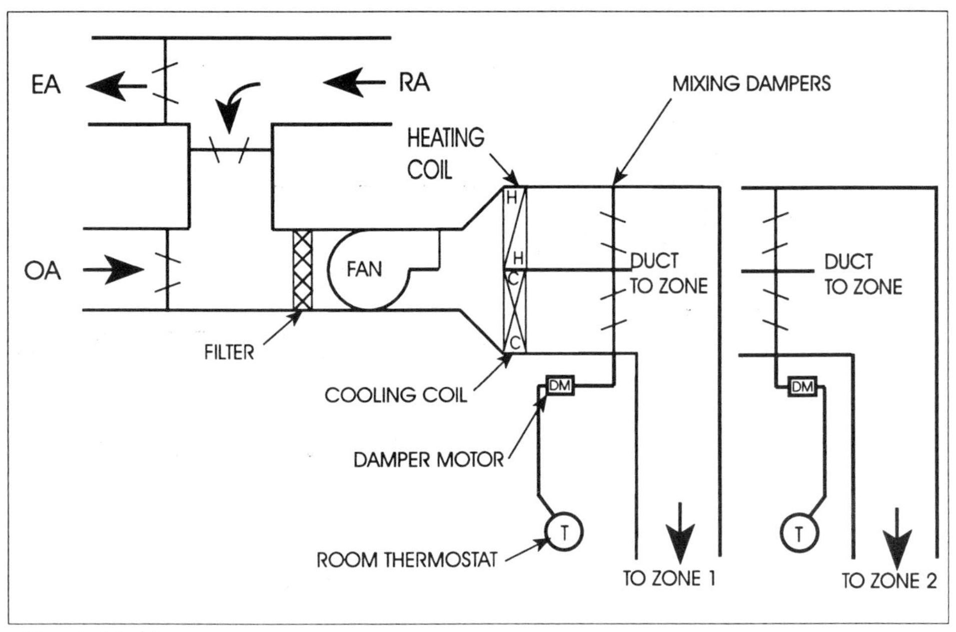

Fig. 4: Multizone system

MULTIZONE SYSTEM

The **multizone system** (Fig. 4) was an early system that was designed to meet the varying needs of different zones. It has a separate supply air duct to each zone in a building. There is a heating coil and a cooling coil in the central air handling unit. Both coils are in operation at the same time. Dampers after the coils mix the hot and the cold supply air to the temperature needed to satisfy each zone.

When heating and cooling occur at the same time it is called **bucking** because the heating and cooling coils are working against each other. The supply air to each zone is mixed to a temperature somewhere in between the hot and the cold supply air. The multizone system uses too much energy to heat and cool the air at the same time.

Because they waste energy, multizone systems are no longer being installed. They are generally banned by local building codes throughout the country.

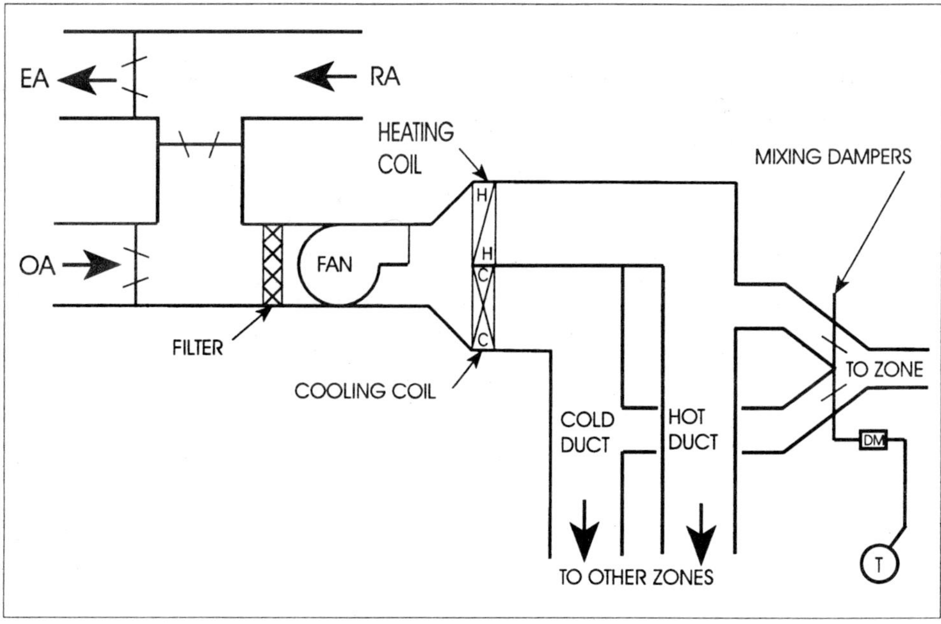

Fig. 5: Dual duct system

DUAL DUCT SYSTEM

The **dual duct low pressure system** (Fig. 5) was also designed to meet the comfort needs of different zones. A dual duct system has two separate supply ducts from the HVAC unit to the outlets in the spaces. One duct supplies cold air, and the other supplies heated air. In this system both the heating and cooling coils operate at the same time, just as with the multizone system. The hot air and the cold air are mixed with dampers **at each zone** in order to obtain the air temperature needed for that zone. This system is intended to be constant volume—variable temperature (CV-VT).

This system also uses too much energy because the hot air and cold air are bucking each other. Therefore the dual duct system that mixes hot and cold air is now generally banned.

The dual duct system also has other problems. The cold duct usually requires most of the supply air. This results in less flow in the hot duct at times and therefore a higher hot duct static pressure. When a zone called for heating, the high static pressure in the hot duct resulted in a high CFM that created drafts and noise in the conditioned spaces.

High Pressure Mixing Boxes

To solve the problems of non-constant airflow rate in a dual duct system, a **high pressure mixing box** was developed. This replaced the mixing dampers. The mixing boxes control the CFM to a constant flow rate. The boxes change the system into a true constant volume—variable temperature (CV-VT).

The dual duct mixing boxes require a high static pressure to operate—usually a minimum of 1.5" wg. The system itself generally requires about 3.0" wg duct pressure to operate properly. The high fan horsepower required to maintain the high static pressure, plus the bucking condition, means a high energy usage. The cost of operating the dual duct system was too high.

Another problem was that, because of the higher pressure often present in the hot duct, the hot air might flow back through the mixing box and into the cold duct. This could raise the air temperature in the cold duct so that the supply air could not cool the spaces adequately.

Low Pressure Reheat Boxes

Later, low pressure **reheat units** for the zones were developed. The supply air had to be cold enough to meet the needs of the zone with the greatest cooling load. The supply air to all other zones had to be reheated. There was no temperature control unless the boiler was operating. In the summer when the boiler was normally turned off, the system could only deliver cold air that was produced by the central cooling system. Often the conditioned spaces were too cold.

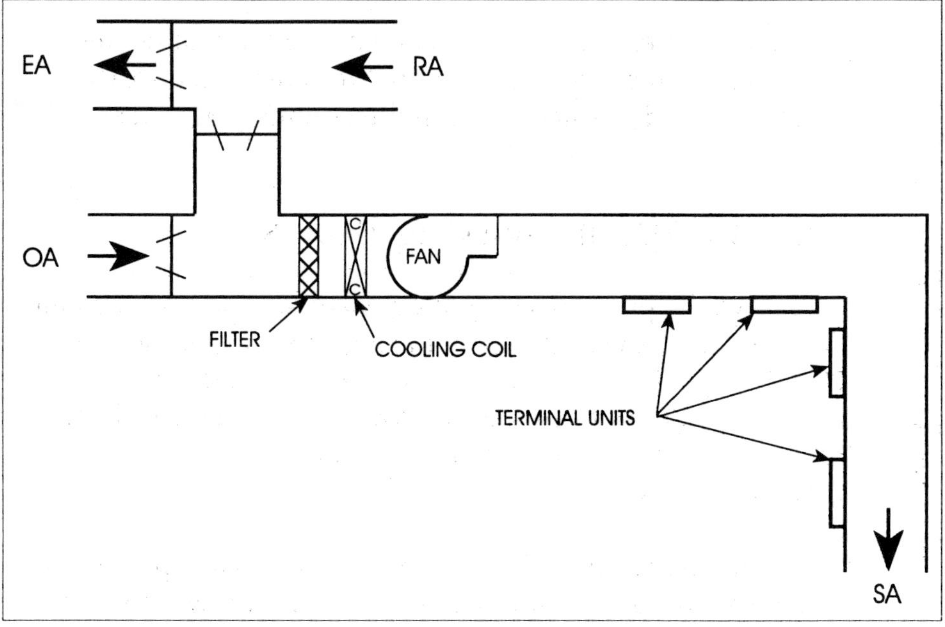

Fig. 6: VAV system

VAV SYSTEMS

Variable air volume (VAV) systems (Fig. 6) were developed to be more energy efficient and to meet the varying heating and cooling needs of the different building zones. A **zone** can be a single room or cluster of rooms with the same heat gain and heat loss characteristics.

VAV systems can save as much as 30% in energy costs as compared to conventional dual duct systems. In addition, they are economical to install and to operate. Duct sizes and central air handling units are smaller and the design and installation is generally much simpler.

The main duct for a typical VAV system provides cooling only (at approximately 55°F). This is called **primary air**. Room thermostats control the amount of primary air delivered to each zone through modulating dampers for each zone. These dampers vary the volume of air to each zone according to the cooling needs.

Early VAV systems varied the fan CFM output according to the total need of the zones. The fan was sized for the maximum probable load. As the air volume for the zones varied, the static pressure (SP) in the main duct tended to vary. An SP sensor in the main duct controlled the fan output to maintain a constant supply duct static pressure. The fan output was varied either by fan inlet vanes or by a damper at the fan outlet. These systems were variable volume—constant temperature (VV-CT)

Early VAV systems were cooling only, so a separate source of heat was needed for the outer rooms. This was usually supplied by perimeter heating in the rooms.

These early VAV systems were low-cost to install. However, depending upon the position of the zone dampers, the zones were subject to delivering too much cold supply air which sometimes created drafts and air noise. These systems were very difficult to balance.

VAV Terminal Units

To better control the secondary air delivered to each zone, the **VAV terminal unit** was developed. A **terminal unit** is a small metal box (Fig. 7) located in the supply air duct just before the outlet of each zone. A terminal unit is also called a **VAV box**, **VAV unit**, or **outlet box**.

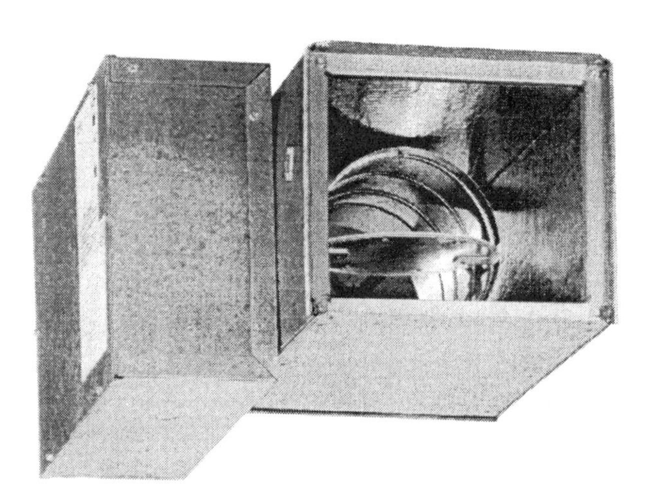

COURTESY OF TITUS

Fig. 7: VAV terminal unit

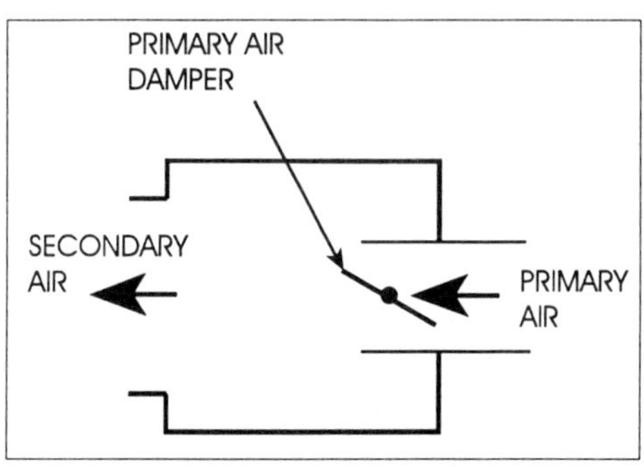

Fig. 8: Simple VAV terminal unit without controls

Each terminal unit (Fig. 8) receives **primary air** from the central air handling unit at the same temperature (about 55°F). The terminal unit contains a primary-air damper (a butterfly damper) which modulates (changes position) according to signals from the automatic control system. (A modulating damper is not just open or closed. It can be at any position in between.) The primary-air damper regulates the volume of cold primary air delivered to the terminal unit according to the needs of the spaces. This is also the volume of **secondary air** delivered to that space. Figure 8 shows only the key features of a terminal unit. Drawings in the units that follow will show how controls and other features allow different types of terminal units to operate.

In its simplest form, the VAV terminal unit provides cooling only. The core spaces generally do not require heating, and the outer rooms are heated by another means. Other terminal units may include a means of heating.

The big advantage of VAV systems with terminal units is that they are able to meet the comfort requirements of different zones in a building without heating and cooling at the same time.

Pressure Dependent or Pressure Independent

VAV systems are either pressure dependent or pressure independent.

The first VAV terminal units were **pressure dependent**. They had no means for limiting the quantity of supply air.

In pressure dependent systems, the volume of air supplied by the terminal unit varies depending upon the **static pressure** (**SP**) in the primary air duct. The primary-air damper in the terminal unit is controlled by a thermostat in the space. However, the airflow through the damper varies according to the SP in the main duct. Terminal units that are close to the supply fan are likely to supply too much primary air. Terminal units that are farthest from the supply fan are not likely to supply enough primary air.

Pressure independent terminal units have flow-sensing devices that limit the flow rate through the box. They can control the maximum and minimum CFM that can be supplied and are therefore independent of the SP in the primary air duct.

Almost all HVAC systems installed or retrofitted at present have pressure independent VAV terminals. Pressure independent systems can be balanced and will allow the correct airflow from each terminal.

Varying Fan Speed

Because each terminal unit regulates its primary air volume independently, the volume (CFM) of primary air delivered by the central air handling unit varies according to the demands of the terminal units in the system. This means that the supply fan in the central air handling unit must vary its output in order to meet the needs of all the terminal units. If the primary-air dampers of most terminal units are full open, the CFM required for the entire system is high. If most terminal unit dampers are closed, the CFM required for the system is much less.

In many current systems, the RPM (speed) of the central supply fan is regulated by the control system to meet the changing demands of the system. A static pressure (SP) sensor in the primary air duct sends a signal to a controller that regulates the fan speed to maintain a constant SP in the primary air duct.

The location of the SP sensor in the primary duct is critical to the performance of the system. It is best placed near the terminal unit that is most difficult to supply. This is the location that has the greatest pressure drop from the fan. If the sensor is placed too close to the supply fan, the SP in the supply duct will be too high during periods of low CFM demand.

IMPROVEMENTS IN VAV SYSTEMS

Early VAV systems were not highly regarded by HVAC technicians. They were considered almost impossible to balance and to keep in balance. Today, pressure independent VAV systems are widely regarded as the best HVAC system design available. This change is largely a result of improvements in the terminal unit.

Many VAV systems and terminal units have been developed to provide for the particular needs of a building. The following commonly used types are described in this book:

- **Cooling-only unit** (Chapter 2)
- **Reheat unit** (Chapter 3) to provide heating as well as cooling
- **Fan-powered reheat unit** (Chapter 4) to provide return airflow through the coil

- **Induction unit** (Chapter 5) using the induction principle rather than a fan to draw return air from the ceiling space
- **Bypass unit** (Chapter 6) to prevent over-cooling by controlling the amount of cold air that enters the space
- **Dual duct system** (Chapter 7) with a hot deck and a cold deck at the air handling unit. The terminal unit selects air from the hot deck or from the cold deck according to the space needs.
- **Changeover-bypass system** (Chapter 8) to provide either heating or cooling with a single supply duct. A bypass damper maintains a constant static pressure in the primary air supply duct.

The seven types of VAV systems and their characteristics are summarized in the following chart:

	Single Duct	Dual Duct	Primary Air		Secondary Air	
			Variable	Constant	Variable	Constant
VAV Cooling-only	■		■		■	
VAV Reheat	■		■		■	
VAV Fan-Powered Reheat	■		■		■	■
VAV Induction	■		■			■
VAV Bypass	■			■	■	
VAV Dual Duct		■	■		■	
Changeover Bypass	■			■	■	

Other Improvements

Digital control systems have greatly improved VAV system performance. Digital controls are more accurate, and they can manage more complex functions. In addition, digital controls input information into a **central processing unit (CPU)**. The CPU is a computer that generates reports analyzing system performance. The CPU can also be used to change the parameters of the system by remote control.

Better sensors are used with the control system to measure such factors as temperature, static pressure, and total pressure more accurately and reliably. This allows for faster and more consistent control of the terminal units.

Better fan speed control allows a VAV system to meet varying demands. Chapter 9 describes different methods of controlling fan speed.

Better design of terminal units has simplified and improved them. Design of outlet diffusers has improved to allow more complete penetration of the conditioned air into the space.

Installation and **balancing methods** have been improved.

CONTROL SYSTEMS

This book does not show all the details of the **automatic control systems**. The diagrams show only enough to provide an understanding of how the terminal unit or the central air handling unit operates. Automatic controls are complex. A complete explanation of them requires much more space. They are covered in other books in this Indoor Environment Technician's Library series.

VAV systems are controlled by either digital or pneumatic control systems.

- **Digital systems** are computer controlled systems that use digital signals.

- **Pneumatic systems** use signals of varying air pressure.

At present, most VAV systems are controlled by digital stand-alone control systems. Digital control systems have many advantages:

- They do not require frequent calibration.
- They can perform complex sequences.
- They can receive instructions from a master computer.
- They can transmit to a master computer information such as damper position, room temperature, supply air quantity, and supply air temperature.

Most of the schematic diagrams in this book are based upon digital systems. Chapter 3, VAV Reheat, shows both digital and pneumatic systems so that you can compare the two.

REVIEW

Write your answers to the following questions without referring to the text. Try to answer every question before you check the answers in the back of the book.

1. How do the needs of the core area of a large building differ from the needs of the rooms on the outer edge of the building?

2. What causes heat loss in a building?

3. What causes heat gain in a building?

4. Why are the multizone systems and dual duct systems no longer used for new buildings?

5. What are two main advantages of a good VAV system?

6. Which is more desirable for a VAV system—a pressure dependent or pressure independent system?

7. What is the reason for sensing static pressure in the primary VAV duct?

8. Which type of controls is most commonly used for new VAV systems—digital or pneumatic?

2 VAV COOLING-ONLY SYSTEM

DESCRIPTION

A VAV cooling-only system is the basic VAV system. Cooling-only units are used mostly for the core spaces where no heating is required. In a large building, the spaces in the core have no outside exposure and therefore no heat loss. They gain heat from occupants and from equipment in the space. Therefore, the core spaces generally require only cooling. The outer spaces either have VAV units that both heat and cool or they have some other heating system.

COURTESY OF TITUS

Terminal unit for a VAV cooling-only system

The central air handling unit must be set so that the minimum amount of outside air required by building codes is always supplied by the VAV terminal units. This provides the ventilation required.

VAV COOLING-ONLY SYSTEM

Single Duct	Dual Duct	Primary Air		Secondary Air	
		Variable	Constant	Variable	Constant

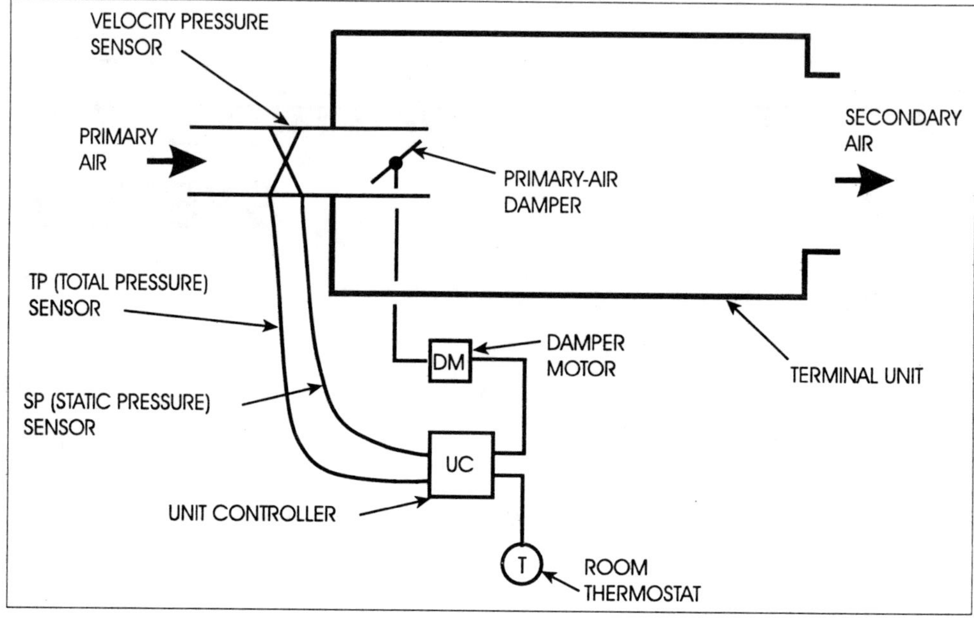

Fig. 1: Cooling-only terminal unit with controls

TERMINAL UNIT

Figure 1 shows the terminal unit for a cooling-only system and the controls associated with it.

- The variable volume primary air is supplied at about 55°F.
- The velocity pressure sensor sends TP (total pressure) and SP (static pressure) signals to the unit controller (UC).
- From these input signals, the UC calculates VP (velocity pressure) and from it the CFM (cubic feet per minute) of the primary air.
- The room temperature sensor (for digital controls), or the room thermostat (for pneumatic controls) also sends a signal to the UC.
- Based upon the input signals, the UC sends a signal to the damper motor (DM).

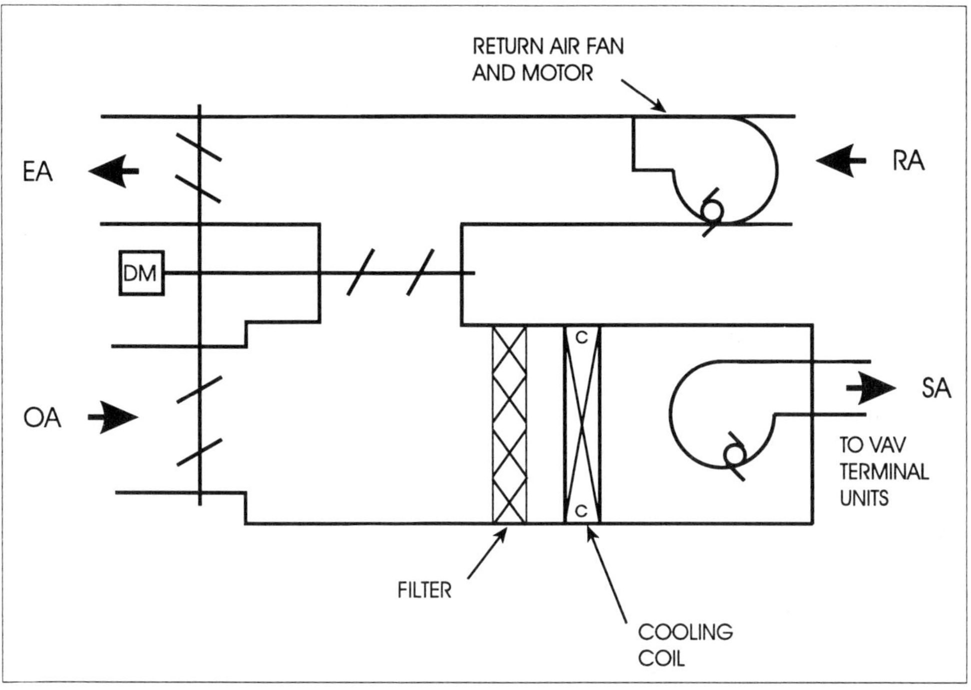

Fig. 2: Central air handling for VAV cooling-only system

- The DM modulates the primary-air damper (a butterfly damper) to vary the primary air quantity, which will equal the quantity of secondary air to the space.

CENTRAL AIR SYSTEM

Figure 2 shows the central air system that supplies primary air to the VAV terminal unit. The following is a typical sequence of the automatic control system.

Unoccupied Mode

- The supply fan and the return fan do not run.
- The OA (outside air) and EA (exhaust air) dampers are closed. This prevents the coil from freezing at low outside temperatures.
- The RA (return air) damper is open.

Morning Cool-Down Cycle

- Cycle starts one to two hours before occupancy.
- Supply and return fans start.
- OA, EA, and RA dampers modulate to supply mixed air cold enough to precool the warmest space.
- Cycle stops when all spaces reach a set temperature.

Occupied Mode

- The occupied time is determined by the automatic control system (similar to a time clock).
- Supply and return fans are on continuously.
- If outside air is above 75°F, the OA damper closes to the minimum position.
- If outside air is below 75°F, the mixed air dampers (OA, EA, and RA) modulate to provide a primary air temperature cold enough to cool the warmest space.
- If the mixed air temperature is too warm, the control system modulates the valve to the chilled water coil to provide the required primary air temperature.
- The primary air discharge temperature is limited to a minimum temperature (approximately 55°F).
- The RPM of the supply air fan is controlled to maintain the required static pressure (SP) in the supply duct.
- The RPM of the return fan is coordinated with the supply fan to maintain a slightly positive building SP (approximately 0.05" wg).

ADVANTAGES OF A COOLING-ONLY SYSTEM

- ❑ The controls are simple.
- ❑ The ductwork can be small because not all zones call for cooling at the same time.

DISADVANTAGES OF A COOLING-ONLY SYSTEM

- ❑ A separate system is needed for heating the outer rooms on the perimeter of the building.
- ❑ Providing the minimum outside air for ventilation may overcool the space.
- ❑ It is often difficult to set a minimum airflow quantity because the VP is so low.

REVIEW

Write your answers to the following questions without referring to the text. Try to answer every question before you check the answers in the back of the book.

1. What rooms in a large building are not likely to need heating? Why?
2. Why are these same rooms likely to need cooling?
3. What is the usual temperature of the primary air for a cooling-only VAV system?
4. Is the quantity of secondary air delivered to the room more or less than the quantity of primary air delivered to the terminal unit?
5. For a VAV system, should the building pressure be slightly negative or slightly positive?

6. What is the purpose of the sensor in the primary air duct just before the terminal unit?

7. About how long before the building is occupied does the cool-down cycle start?

8. When do the supply fan and return fan operate?

9. What damper setting must conform to building codes?

10. What are two kinds of information sent to the unit controller?

11. Under what conditions does the OA (outside air) damper close to the minimum position?

12. On the drawing of the central air handling system, draw the RA, OA, and EA dampers to show their position when the building is not occupied.

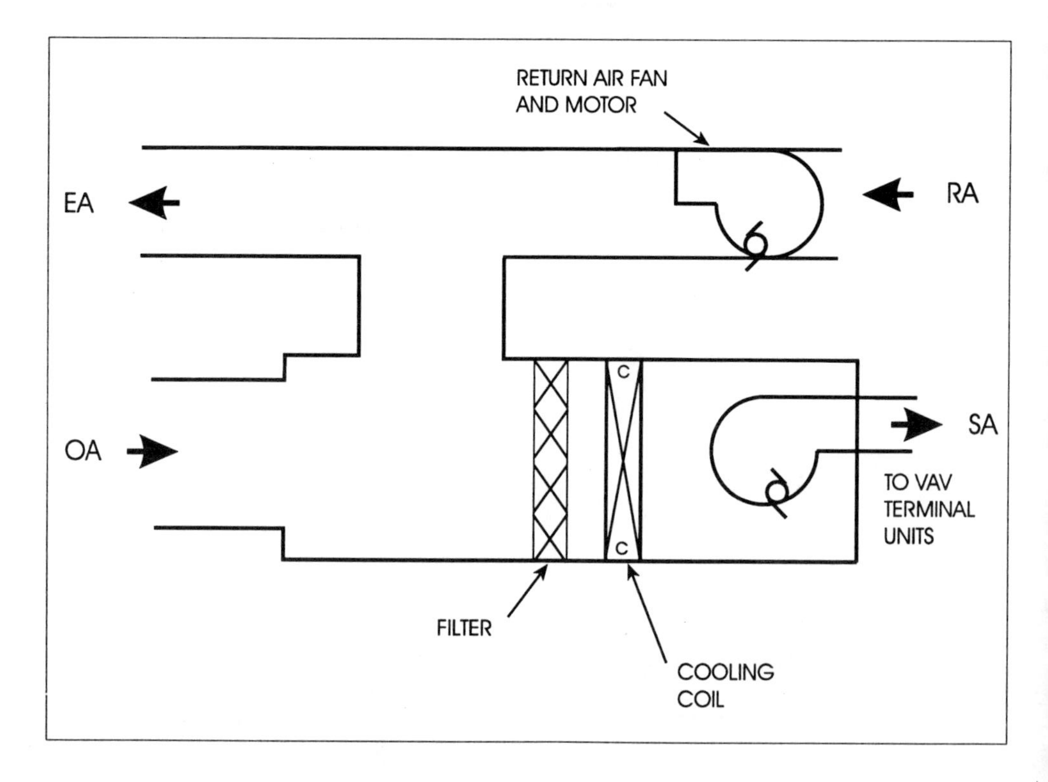

3

VAV REHEAT SYSTEM

DESCRIPTION

The **VAV reheat system** can either heat or cool the space which it serves. It is like the **VAV cooling-only** system in Chapter 2 except that a reheat coil has been added to the terminal unit. VAV reheat units are used for the outer rooms where heating as well as cooling is needed. VAV cooling-only units are used mostly for the core spaces where no heating is required.

For all systems, the central air handling unit must be controlled so that the minimum amount of outside air required by building codes is always supplied to all zones of the building. This provides the ventilation required.

VAV REHEAT SYSTEM

Single Duct	Dual Duct	Primary Air		Secondary Air	
		Variable	Constant	Variable	Constant

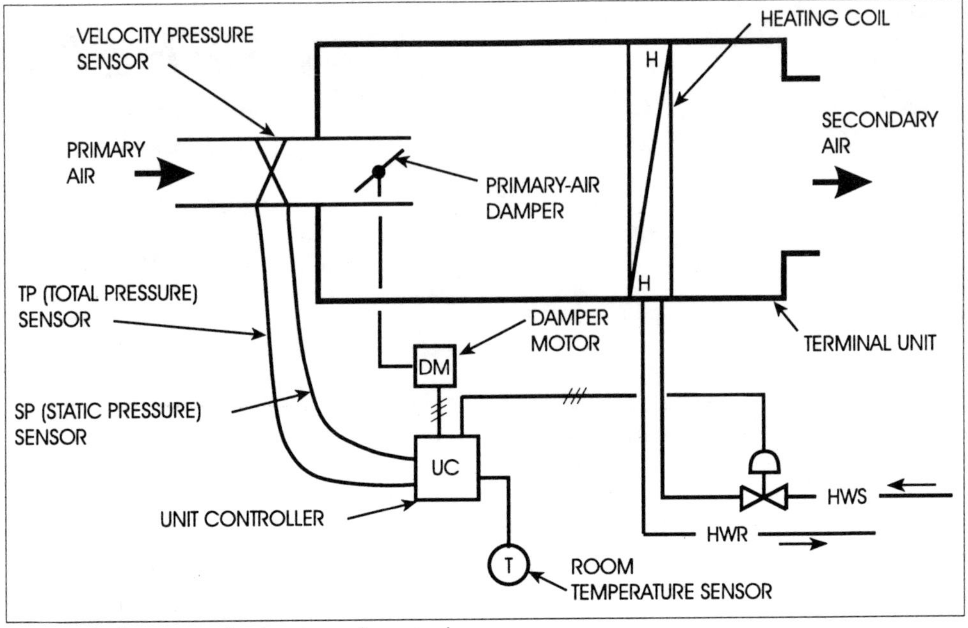

Fig. 1: Reheat terminal unit and controls

TERMINAL UNIT

Figure 1 shows the terminal unit for a VAV reheat system and the controls associated with it.

- The variable volume primary air is supplied at about 55°F.
- The velocity pressure sensor sends TP (total pressure) and SP (static pressure) signals to the unit controller (UC).
- The room temperature sensor (for digital controls) or the room thermostat (for pneumatic controls) also sends a signal to the UC.
- Based upon the input signals, the UC sends signals to the damper motor (DM) **and also to the reheat coil.**
- The automatic valve on the HWS (hot water supply) modulates according to the need for heat.

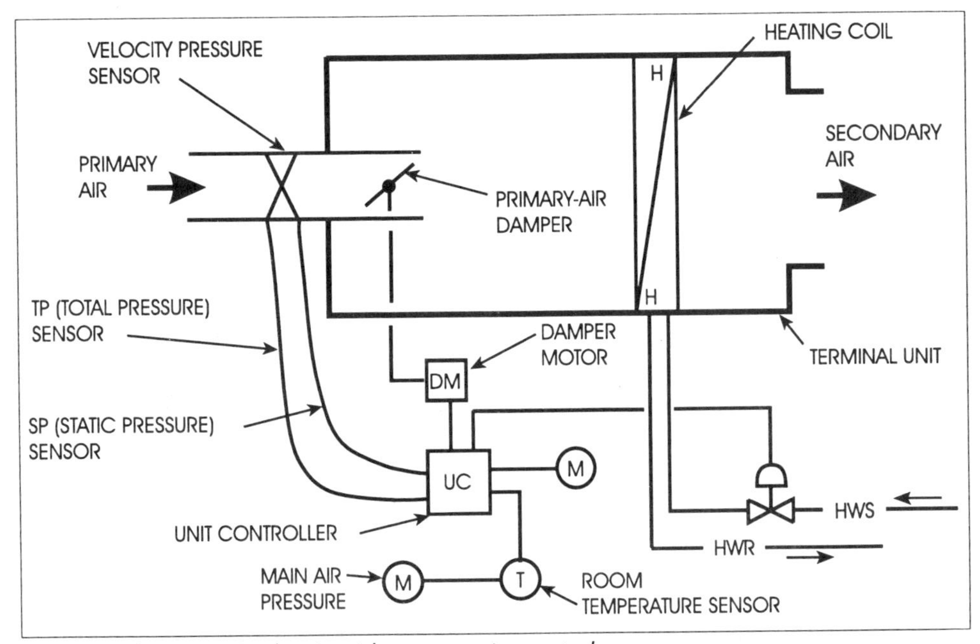

Fig. 2: Reheat terminal unit with pneumatic controls

- The temperature sensor set point is generally about 70°F for heating and 75°F for cooling.
- When the space calls for heat, the primary-air damper closes down to a minimum airflow for the terminal unit in the heating mode (about 50% of the maximum flow). The secondary air delivered to the space is constant at a minimum value during the heating cycle.
- When the space temperature rises above the set point, the terminal unit returns to the cooling mode (heat valve closed) and the primary air quantity varies to meet the cooling load.

Pneumatic Controls

Today most VAV systems are controlled by digital controls. They provide data that can be stored and retrieved for reports and for additional control processes. However, some systems are controlled by pneumatic systems. Figure 2 shows the terminal unit reheat controlled by a pneumatic system.

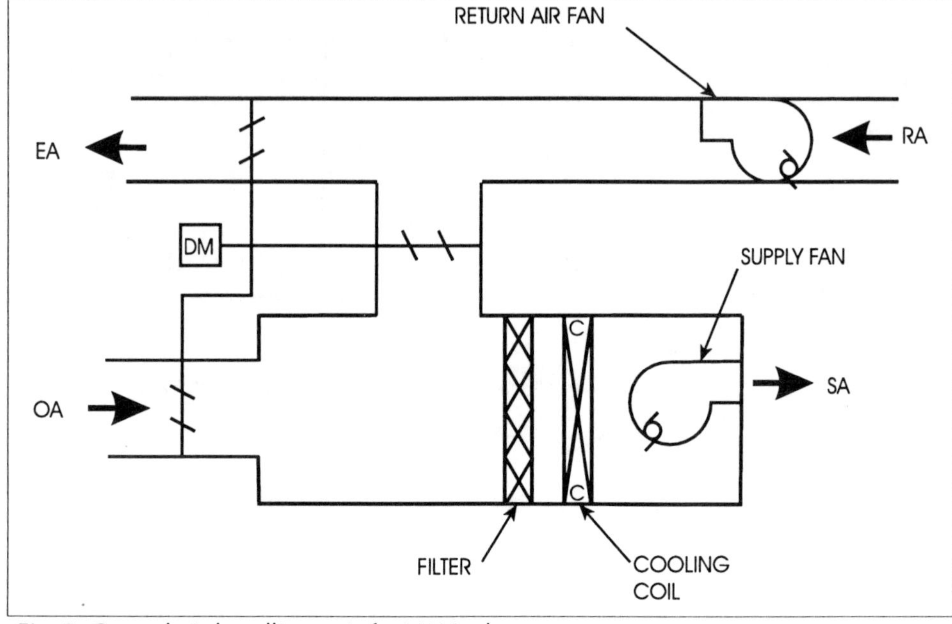

Fig. 3: Central air handling unit for VAV reheat system

CENTRAL AIR SYSTEM

The central air handling system (Fig. 3) is the same as that for a VAV cooling-only system. Primary air at about 55°F is supplied to all terminal units. The following is a typical sequence of the control system.

Unoccupied Mode

- The supply fan and the return fan do not run.
- The OA (outside air) and EA (exhaust air) dampers are closed. This prevents the coil from freezing at low outside temperatures.
- The RA (return air) damper is open.
- If the temperature in a space falls below a set temperature (usually about 60°F), the supply and return air fans start. The OA and EA dampers

remain closed. The rooms are heated by the reheat terminal units.

- The fans run until the coldest room is satisfied (reaches 60°).

Morning Warm-Up Cycle

- Cycle starts one to two hours before occupancy.
- The central supply and return air fans start.
- OA and EA dampers remain closed.
- Reheat coils in terminal units are activated.
- Cycle stops when all spaces reach the set temperature.

Morning Cool-Down Cycle

- Cycle starts one to two hours before occupancy.
- Supply and return air fans start.
- OA, EA, and RA dampers modulate to supply mixed air cold enough to precool the warmest space.
- Cycle stops when all spaces reach a set temperature.

Occupied Mode

- The occupied time is determined by the control system.
- Supply and return fans are on continuously.
- If outside air is above 75°F, the OA damper closes to the minimum position.
- If outside air is below 75°F, the mixed air dampers (OA, EA, and RA) provide the primary air duct with the air temperature that will cool the warmest space.

- If the mixed air temperature is too warm, the control system will open the valve to the chilled water coil to provide the required primary air temperature.
- The primary air discharge temperature will be limited to a minimum temperature (approximately 55°F).
- The RPM of the supply air fan is controlled to maintain the required static pressure (SP) in the duct.
- The RPM of the return fan is coordinated with the supply fan to maintain a slightly positive building SP (static pressure) (approximately 0.05" wg).

ADVANTAGE OF A REHEAT SYSTEM

- No separate heating system is needed.

DISADVANTAGES OF A REHEAT SYSTEM

- When heating is called for, energy is wasted because the primary air is cooled and then reheated in the terminal unit. However, the primary air temperatures are kept only as cool as necessary to satisfy the warmest zone.
- Hot water supply and return lines for the reheat coil can leak. They should be run above the ceiling in the hallways rather than in the rooms. If there are any leaks in the piping, removable ceiling panels in the hallway make repairs easier.

REVIEW

Write your answers to the following questions without referring to the text. Try to answer every question before you check the answers in the back of the book.

1. For the VAV reheat unit, what input information is sent to the terminal unit controller?
2. Why does the primary-air damper remain partially open when the space is calling for heat?
3. What modulates to provide the amount of heat needed?
4. If the temperature in some of the spaces falls below 60°F while the building is unoccupied, how does the central air system respond?
5. During what period do the supply and return fans operate continuously?
6. Are the EA and OA dampers closed during the morning cool-down or during the morning warm-up cycle?
7. If the outside air is 85°F, what is the position of the OA damper?
8. Is there bucking in this system?

9. On the drawing, show the RA, OA, and EA dampers as they would be during the occupied mode if the space is calling for cooling and the outside temperature is 68°F.

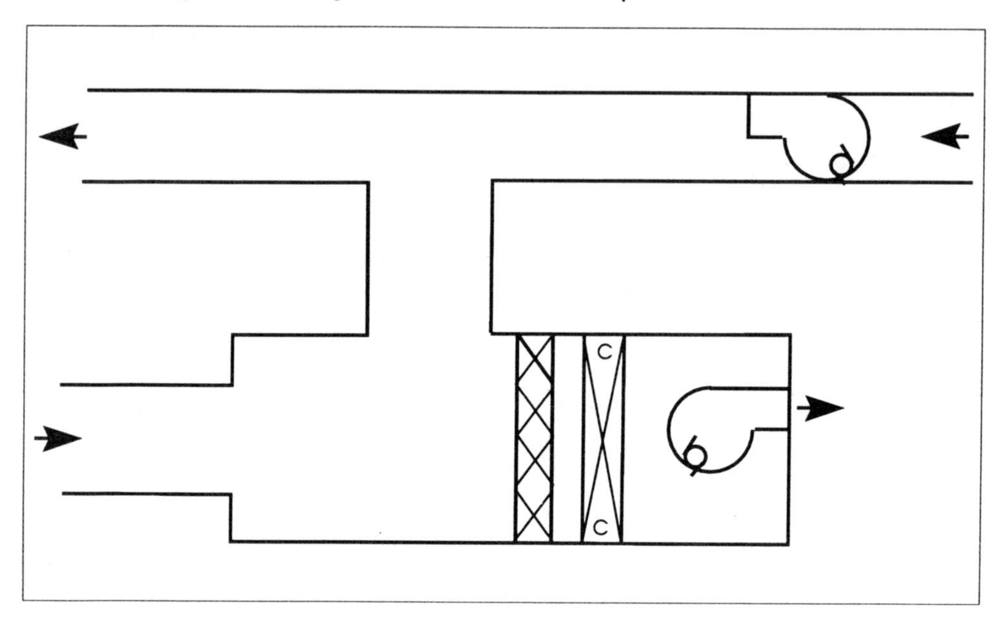

10. Connect the damper motor and the hot water supply valve to the VAV system.

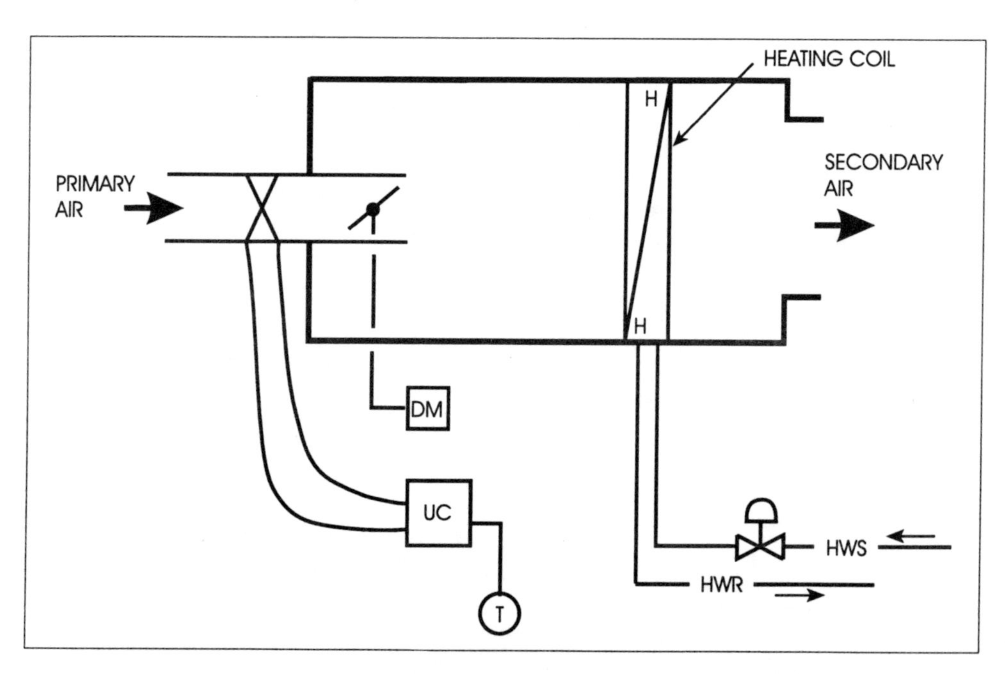

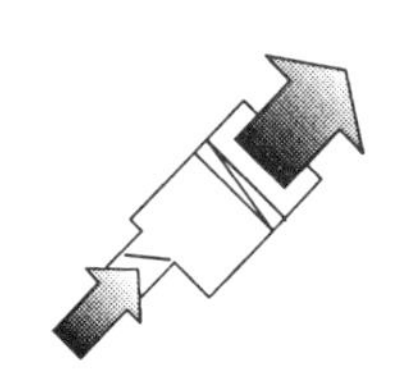

VAV FAN-POWERED REHEAT SYSTEM

DESCRIPTION

The **VAV fan-powered reheat system** is similar to the VAV reheat system described in Chapter 3. The difference is that a fan is added to overcome the resistance through the coil in the terminal unit. (The reheat unit without a fan uses only the pressure of the primary air to move the air through the coil.) The terminal unit fan motor has a **variable speed controller** which can be manually set for the required speed.

A fan-powered reheat system may use an **electric resistance heater** instead of a hydronic heating coil. An electric heater avoids the problem of possible leaks above the ceiling.

There are two types of terminal units for fan-powered reheat:

- ❑ Series fan
- ❑ Parallel fan

In the heating mode, the **series fan** unit provides constant secondary air. The **parallel fan** unit provides variable secondary air delivered to the space.

VAV FAN-POWERED REHEAT SYSTEM

Single Duct	Dual Duct	Primary Air		Secondary Air	
		Variable	Constant	Variable	Constant

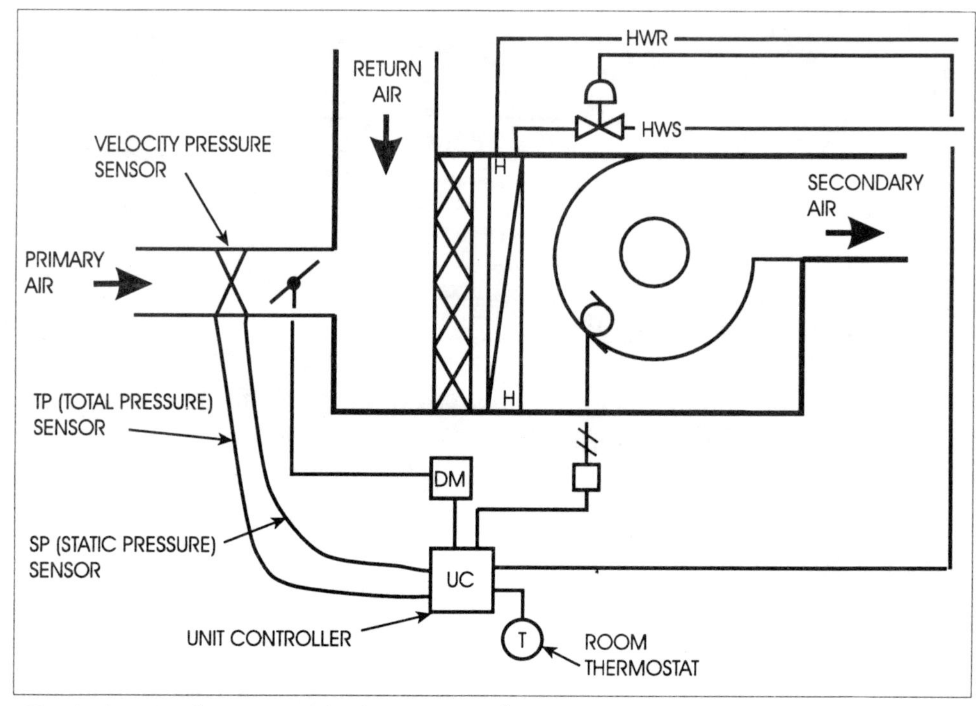

Fig. 1: A series fan-powered reheat terminal unit

SERIES FAN-POWERED TERMINAL UNIT

Figure 1 shows a **series fan terminal unit**. The fan and coil are in series with the primary air supply fan. This is called a **series fan** because the primary air must flow through the fan. The primary air and secondary air are in series.

The terminal unit fan runs constantly. Its volume must be greater than the primary air volume in order to keep a negative pressure so that no air flows out through the return air inlet.

This is a **variable primary—constant secondary** system. The primary airflow is **variable**. The secondary airflow is **constant**. The following is a typical sequence of the control system.

Unoccupied Mode

- ❑ The terminal unit fan runs with the central air fan off to maintain a minimum set temperature in the space. When the space is satisfied, the terminal unit fan shuts off and the heating coil is deactivated.

Occupied Mode

- ❑ The terminal unit fan and the central air fan run constantly when the space is occupied, whether the unit is supplying cooled or warm air.
- ❑ The fans are generally controlled by a time clock that is set for the times of space occupancy.
- ❑ Sometimes an occupancy sensor is used in place of a time clock. The occupancy sensor senses movement in the room and starts the terminal unit fan.

Heating while Occupied

- ❑ The terminal unit fan is on.
- ❑ The primary air damper is at a minimum position.
- ❑ The reheat coil valve modulates according to the signal from the space temperature sensor.

Cooling while Occupied

- ❑ The terminal unit fan is on.
- ❑ The reheat coil is inactive.
- ❑ The primary air valve modulates.
- ❑ The fan supplies mixed primary air and return air to the space. The mixed air is mostly primary air.

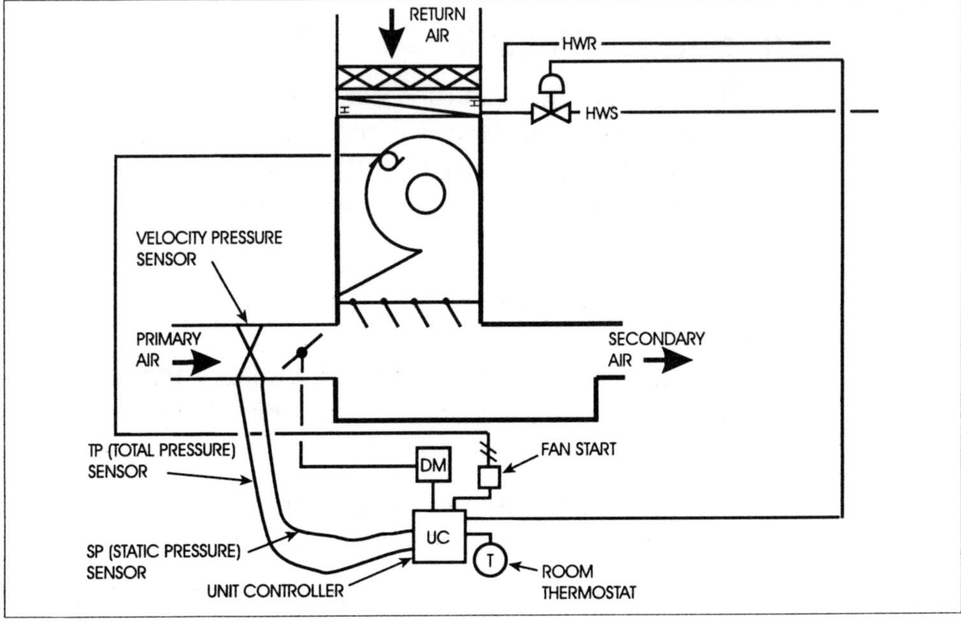

Fig. 2: A parallel fan-powered reheat terminal unit

PARALLEL FAN-POWERED TERMINAL UNIT

Figure 2 shows a **parallel fan-powered terminal unit**. This is a **variable primary—variable secondary** system. In the heating mode, both the primary air quantity and the secondary air quantity are **variable**.

This system is called **parallel** because the primary air is in a separate path from the return air, fan, and coil (Fig. 2). The fan is at a right angle to the primary air to provide a good mix of the air from the two ducts.

The terminal unit fan takes return air directly from the air above the ceiling. A **backdraft damper** at the fan prevents primary air from flowing backwards through the fan and into the ceiling space when the terminal unit fan is off.

Because the secondary air is supplied from two sources, this type of terminal unit is difficult to balance, especially if the cold and hot air streams do not mix in the terminal unit.

The following is a typical sequence of the control system.

Unoccupied Mode

- No primary air.
- Central system fans are off.
- If the temperature in the space falls below a set level (usually about 60°F), the unit fan and reheat coil operate. Otherwise, the fan is off and the heating coil is inactive.

Morning Cool-down Cycle

- Cycle starts one to two hours before occupancy.
- Central system supply and return air fans start.
- OA, EA, and RA dampers modulate to supply mixed air cold enough to precool the warmest space.
- Cycle stops when all spaces reach a set temperature.

Cooling

- The terminal unit operates like a VAV cooling-only terminal unit.
- The terminal unit fan is inactive.
- The reheat coil is inactive.
- The primary air damper modulates from minimum to maximum according to space need.

Heating

- The terminal unit fan operates only on a call for heat.
- The hot water supply valve modulates according to a signal from the temperature sensor in the space.
- Primary air damper is at a minimum position.

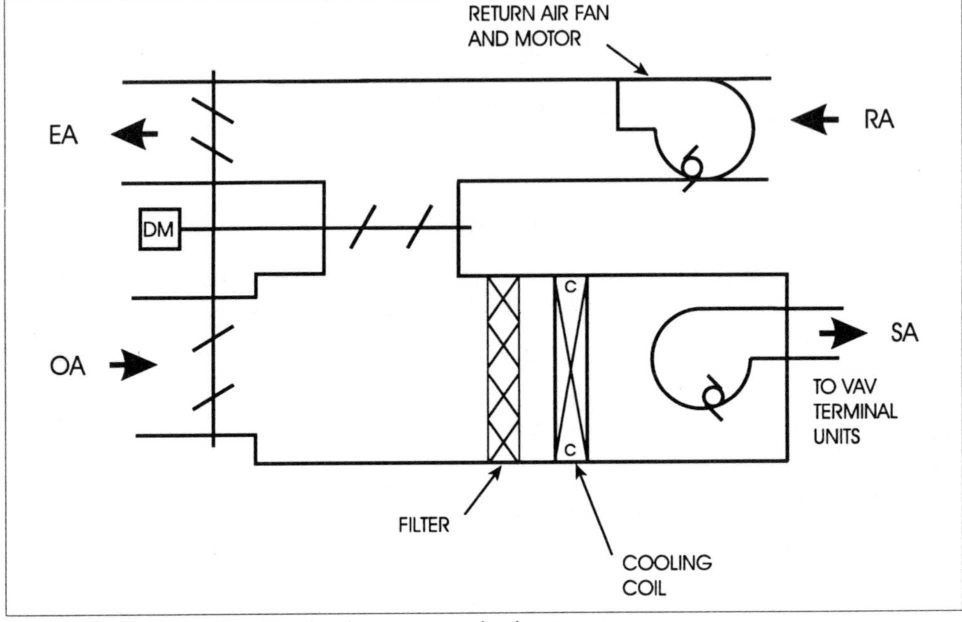

Fig. 3: Central air handler for fan-powered reheat system

CENTRAL AIR

The central air system (Fig. 3) provides a variable flow of primary air. The fan output is controlled by a static pressure sensor in the main duct (as it is for the VAV cooling-only system).

Unoccupied Mode

- The central supply fan and the return fan do not run.
- The OA (outside air) and EA (exhaust air) dampers are full closed. This prevents the coil from freezing at low outside temperatures.
- The RA (return air) damper is open.
- If the temperature in a space falls below a set temperature (usually about 60°F), the terminal unit fan and reheat coil operate, but the central system fans remain off.

Morning Warm-up Cycle

- The central air system does not operate.
- The terminal unit fans cycle on and off with the reheat coil activated to satisfy the space requirements.

Morning Cool-down Cycle

- Cycle starts one to two hours before occupancy.
- The central supply and return fans start.
- Terminal unit fans are off.
- OA, EA, and RA dampers modulate to supply mixed air cold enough to precool the warmest space.
- Cycle stops when all spaces reach a set temperature.

Occupied Mode

- The hours of occupied time are preset by clock.
- Supply and return fans are on continuously.
- If outside air is above 75°F, the OA damper modulates to the minimum position.
- If outside air is below 75°F, the mixed air dampers (OA, EA, and RA) provide the primary air duct with the air temperature that will cool the warmest space.
- If the mixed air temperature is too warm, the control system opens the valve to the chilled water coil to provide the required primary air temperature.
- The primary air discharge temperature is limited to a minimum temperature (usually around 55°F)
- The RPM of the supply air fan is controlled to maintain the required SP (static pressure) in the duct.
- The RPM of the return fan coordinates with the supply fan to maintain a slightly positive building SP (usually around 0.06" wg).

ADVANTAGES OF A FAN-POWERED REHEAT SYSTEM

- ❑ No separate heating system is needed because of the reheat coil.
- ❑ The fan provides better control of the airflow to the spaces.

DISADVANTAGES OF A FAN-POWERED REHEAT SYSTEM

- ❑ The fan in the terminal unit may create undesirable noise.
- ❑ The hot water supply and return for the reheat coil may cause leaks. An electric resistance heater avoids this problem
- ❑ The fan and motor in the terminal unit may require maintenance and replacement. It may be difficult to reach the fan through ceiling access panels.
- ❑ Dirt from the ceiling space can foul the reheat coil, so a filter should be installed in front of the coil. The filter and coil may be difficult to reach, so they may not be maintained properly.
- ❑ In the parallel unit, the cold and hot air streams may not mix completely. As a result, one outlet for the terminal may receive mostly cold air while another may receive mostly hot air. Deflectors may be added to create air turbulence to mix the two streams.
- ❑ Because the secondary air is supplied from two sources, the parallel unit is difficult to balance, especially if the cold and hot air streams do not mix in the terminal unit.

REVIEW

Write your answers to the following questions without referring to the text. Try to answer every question before you check the answers in the back of the book.

1. With fan-powered reheat units, does the central system operate during the morning warm-up cycle?

2. With fan-powered reheat units, does the central system operate during the morning cool-down cycle?

3. Why are the OA and EA dampers usually full closed when the building is unoccupied?

4. What is the advantage of having an electric resistance heater rather than a hot water coil in the terminal unit?

5. What is the disadvantage of an electric resistance heater?

6. Does the terminal unit in the drawing below have a series or a parallel fan?

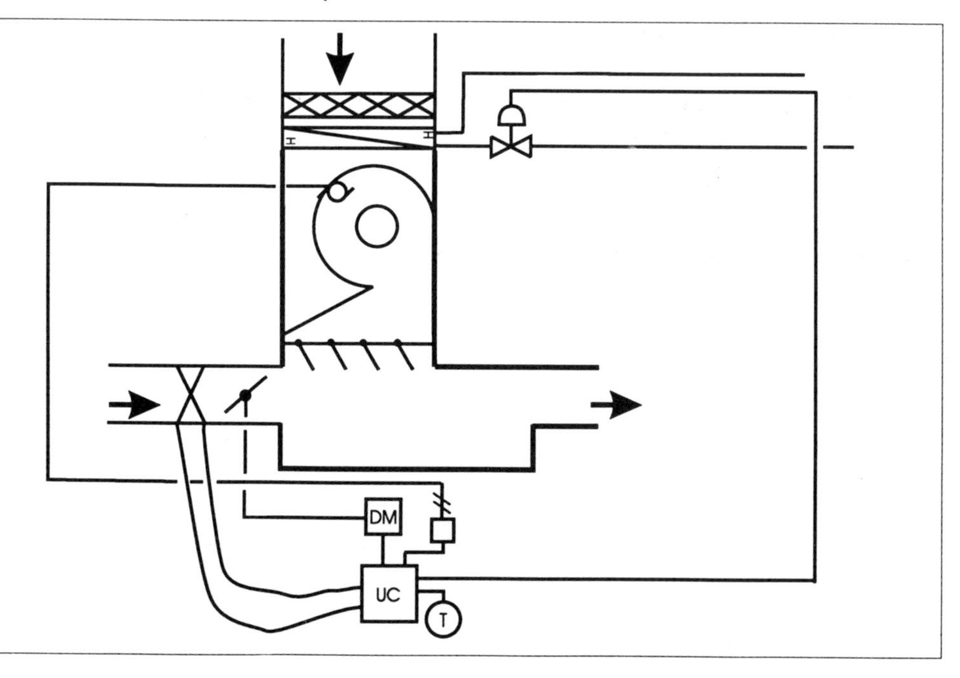

7. Does the terminal unit in the drawing below have a series or a parallel fan?

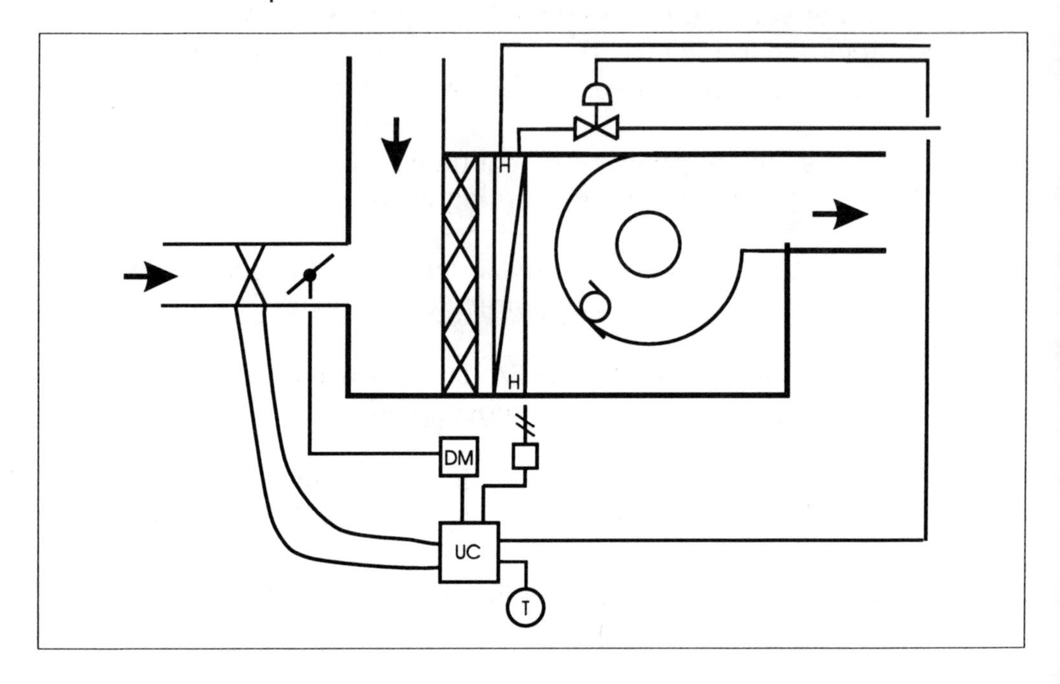

8. Which type of unit—series or parallel—requires a backdraft damper?

9. Does a parallel fan in the terminal unit operate when the space is calling for cooling or when the space is calling for heating?

10. Does a series fan in the terminal unit operate when the space is calling for cooling or when the space is calling for heating?

5

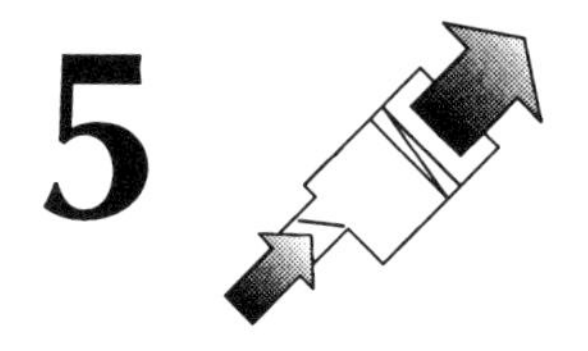

VAV INDUCTION SYSTEM

DESCRIPTION

The purpose of the **VAV induction unit** is to provide a constant supply of secondary air without a fan in the terminal unit. It can be used with or without a reheat coil. This system is currently available but is not a popular system. It is usually encountered in older buildings.

The induction unit works by supplying a stream of high-pressure primary air to the terminal unit. This causes return air from above the ceiling to be induced into the unit. As the pressure in the unit after the primary damper varies, it causes the amount of return air induced into the terminal to vary. This maintains a relatively constant flow of air being supplied to the space.

The induction unit shown in Fig. 1 has pneumatic controls.

VAV INDUCTION UNIT SYSTEM

Single Duct	Dual Duct	Primary Air		Secondary Air	
		Variable	Constant	Variable	Constant

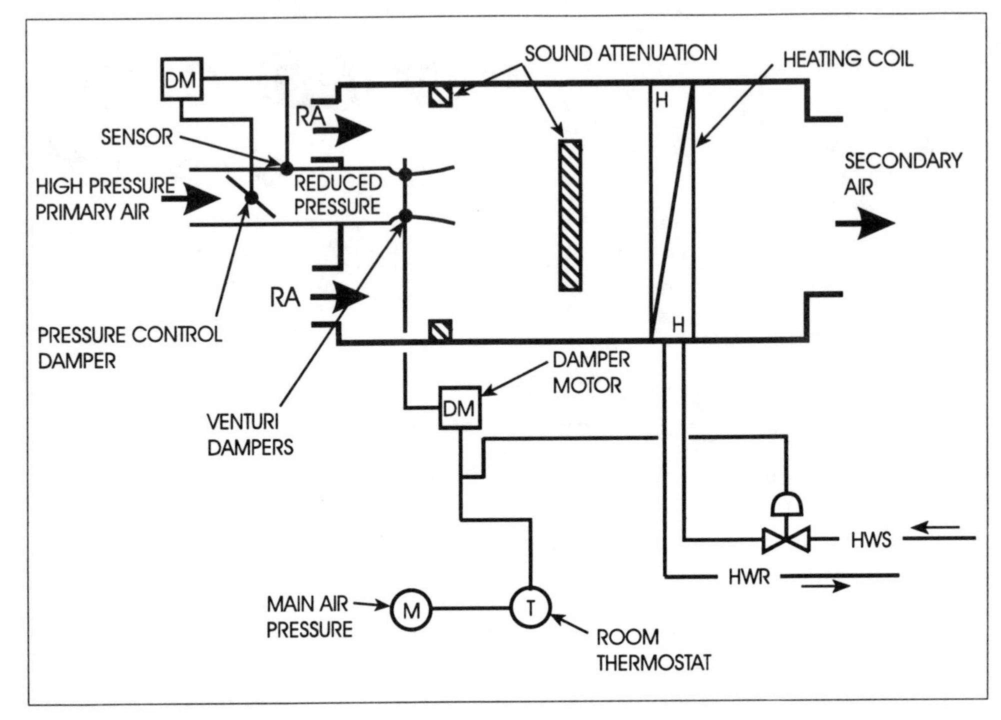

Fig. 1: Induction terminal unit with pneumatic controls

TERMINAL UNIT

The **primary-air damper** for the induction unit is venturi-shaped (Fig. 1). This opening is similar to the venturi used in auto carburetors and in venturi flow meters. As the air flows out of the dampers, it creates a negative pressure outside the opening. This induces return air in the ceiling spaces to enter the terminal unit and mix with the primary air. The primary air always flows and is at least 25% of the secondary airflow quantity. This keeps it from flowing back through the return air inlets.

The damper motor modulates the venturi damper according to a signal from the temperature sensor in the space. The damper modulates to vary the quantity of primary air entering the terminal unit. As a result, it also controls the quantity of return air entering the terminal unit:

- As the damper closes, the primary air quantity is reduced but the air velocity is increased. The greater air velocity creates a lower negative pressure within the unit which induces more return air.
- As the damper opens, the air quantity is increased but the air velocity is reduced. The lower air velocity creates less negative pressure so less return air is induced into the space.

The primary airflow CFM must always be equal to at least 25% of the maximum primary air quantity, or the venturi action will not take place.

The primary air just before it enters the terminal unit must be maintained at a constant SP (static pressure). A static pressure sensor in the primary duct at the unit modulates a **pressure control damper** to maintain a constant SP entering the unit.

The induction system also requires high pressure in the primary air duct. It is generally between 1.5" wg and 2.0" wg.

The following is a typical sequence of the control system.

Heating Mode

- The damper motor modulates the primary-air damper to a minimum position so that only minimum primary air is supplied to the terminal unit.
- The increased air velocity at the venturi induces more return air into the unit and discharges a relatively constant total airflow quantity through the reheat coil.
- The HWS (hot water supply) valve modulates to provide hot water to the reheat coil.

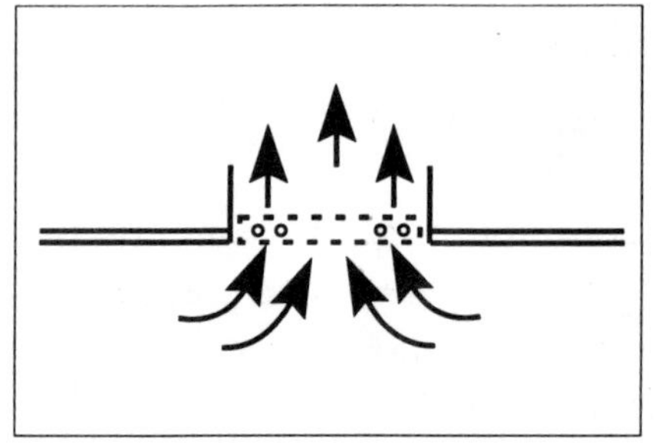

Fig. 2: A heat-of-light return air fluorescent fixture adds some heat to return air

- ❑ If there is no reheat coil, the return air may be drawn through fluorescent heat-of-light fixtures (Fig. 2) in the ceiling. The heat from the lights provides some heat to the return air space above the ceiling.

Cooling Mode

- ❑ The HWS valve closes so that there is no heat in the reheat coil.
- ❑ The damper motor modulates the primary-air venturi damper as required.
- ❑ As the damper opens more fully, less return air is induced. This arrangement maintains a nearly constant airflow to the space.

CENTRAL AIR SYSTEM

The central air system (Fig. 3) is similar to that for the VAV cooling-only system. The difference is that the fan is designed for higher static pressure in the supply air duct (1.5" wg to 2.0" wg at the most distant induction unit).

Unoccupied Mode

- ❑ The supply fan and the return fan do not run.
- ❑ The OA (outside air) and EA (exhaust air) dampers are closed. This prevents the coil from freezing at low outside temperatures.
- ❑ The RA (return air) damper is open.

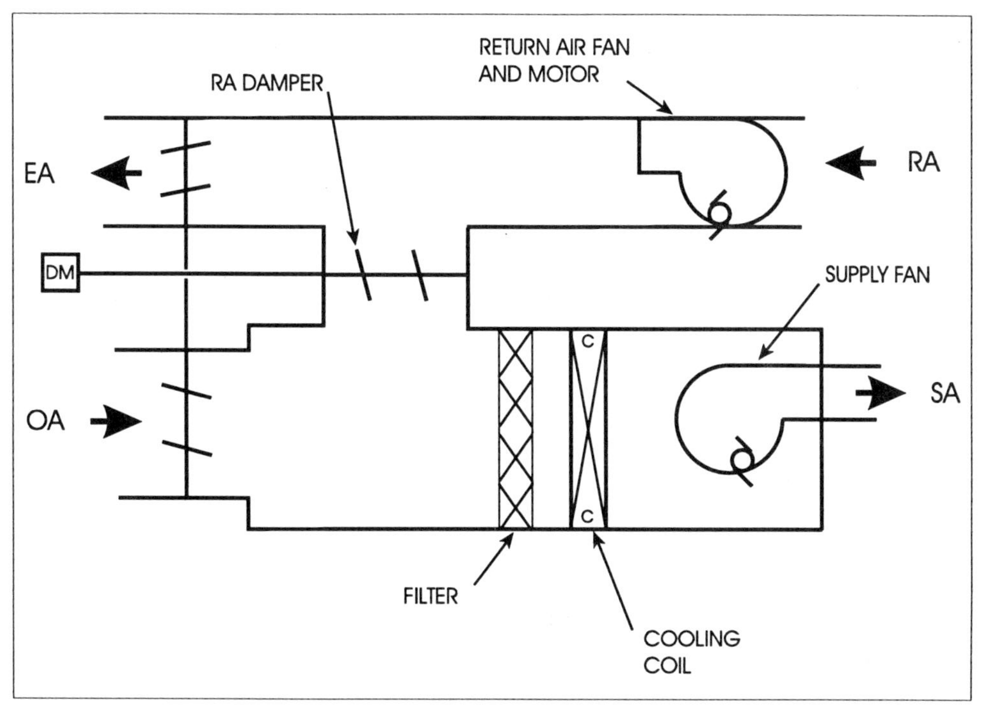

Fig. 3: Central air handler for VAV induction system

- ❑ If the temperature in a space falls below a set temperature (usually about 60°F), the supply and return air fans start. The OA and EA dampers remain closed.
- ❑ The fans run until the coldest room is satisfied (reaches about 60°F).

Morning Warm-Up Cycle

- ❑ The cycle starts one to two hours before occupancy.
- ❑ The supply and return air fans start.
- ❑ OA and EA dampers remain closed.
- ❑ Cycle stops when all spaces reach the set temperature.

- ❑ If the unit has a heating coil, the valves modulate according to the needs of the space.

Morning Cool-Down Cycle

- ❑ The cycle starts one to two hours before occupancy.
- ❑ Supply and return fans start.
- ❑ OA, EA, and RA dampers modulate to supply mixed air cold enough to precool the warmest space.
- ❑ The cycle stops when all spaces reach a set temperature.

Occupied Mode

- ❑ The control system sets the hours of occupied time.
- ❑ Supply and return fans are on continuously.
- ❑ If outside air is above 75°F, the OA damper modulates to the minimum position.
- ❑ If outside air is below 75°F, the mixed air dampers (OA, EA, and RA) modulate to provide the primary air temperature that will cool the warmest space.
- ❑ If the mixed air temperature is too warm, the control system opens the valve to the chilled water coil to provide the required primary air temperature.
- ❑ The primary air discharge temperature is limited to a minimum temperature (usually around 58°F).
- ❑ The RPM of the supply air fan is controlled to maintain the required static pressure (SP) in the duct.

The RPM of the return fan coordinates with the supply fan to maintain a slightly positive building SP (usually around 0.05" wg).

ADVANTAGES OF AN INDUCTION SYSTEM

- ❑ The airflow to the space is almost constant. This makes the space more comfortable.
- ❑ There are no motors above the ceiling to maintain.

DISADVANTAGES OF AN INDUCTION SYSTEM

- ❑ More fan horsepower is required to provide the high pressure primary air.
- ❑ The high horsepower fan for the primary air is likely to be noisy.
- ❑ The ductwork must be constructed to high pressure standards.
- ❑ Cool primary air must be supplied in order to get warmer return air.
- ❑ If the unit has a hydronic reheat coil, there is a possibility of leaks.

REVIEW

1. Is the secondary air in the system variable or constant?
2. What is the purpose of a venturi damper?
3. At what point should the primary air pressure be constant?
4. When the space calls for heating, is the primary-air damper for the terminal unit in the minimum or the maximum position?

5. Does the central air unit provide heating or cooling?

6. Is the primary air pressure kept fairly high or fairly low?

7. If the venturi opens to allow more primary air to enter the unit, how is the return air affected?

6 VAV BYPASS SYSTEM

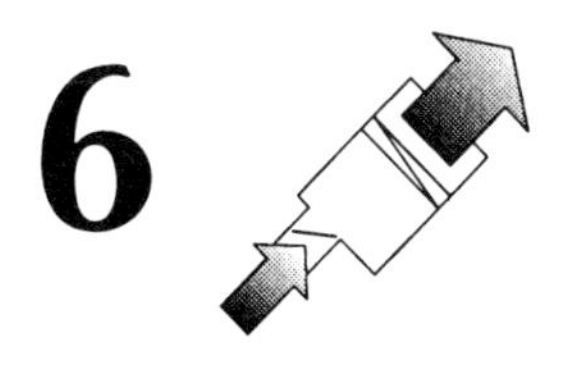

DESCRIPTION

The VAV bypass system is cooling only. It is used for smaller systems. The VAV bypass system has the advantage of a relatively low installation cost. However, it has a relatively high energy consumption. This system is not currently popular but is still available. It is usually found in older buildings.

There is no primary-air damper for this system. Instead, a damper modulates to control the secondary supply air outlet and the return air outlet. When more air goes to the secondary supply air, less goes to the return air path (generally above the ceiling). When less goes to the secondary supply air, more goes to the return air.

A bypass system must have a return air fan in the central air system to prevent too much air pressure in the ceiling.

VAV BYPASS SYSTEM

Single Duct	Dual Duct	Primary Air		Secondary Air	
		Variable	Constant	Variable	Constant

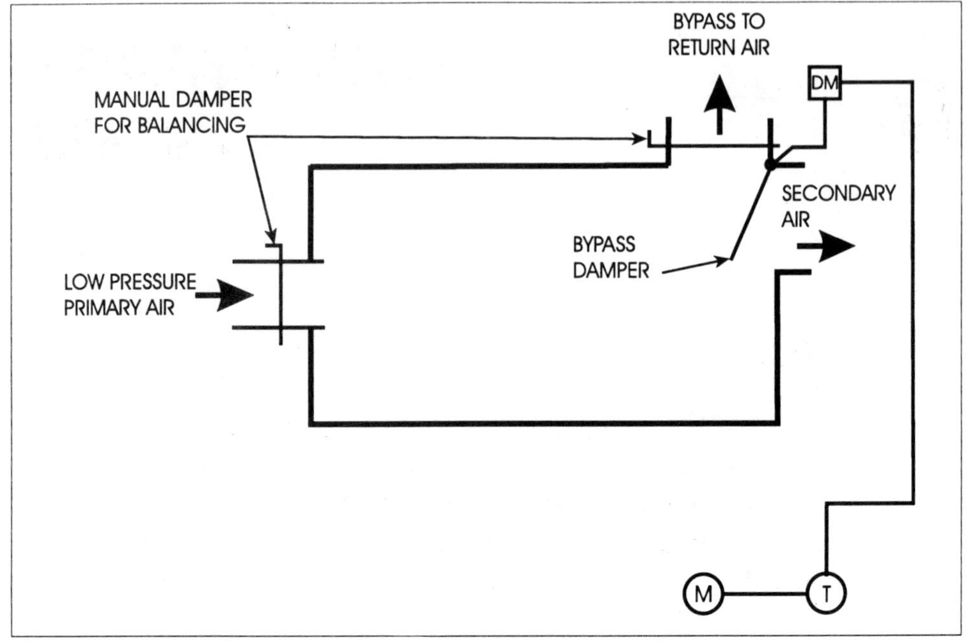

Fig. 1: Bypass terminal unit

TERMINAL UNIT

The primary air is cooled to about 55°F. There is no operator on the primary air valve and no velocity sensor (Fig. 1). This means that the primary air is not controlled at the unit. Instead, a motorized damper controls the amount of primary air entering the secondary duct. Primary air that is not supplied to the space enters the return air ceiling plenum and returns to the central air handler.

The following is a typical sequence of the control system.

Cooling

- ❐ The damper motor modulates the secondary air damper according to a signal from the thermostat in the space.

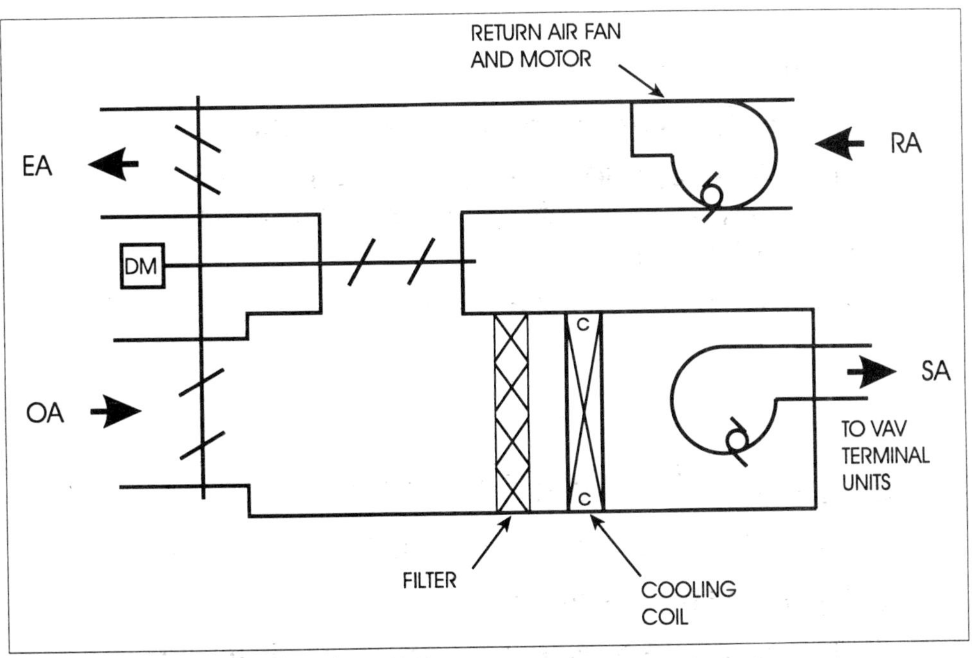

Fig. 2: Central air handling for VAV bypass system

- Cooling is supplied by the coil in the central air system.
- For full cooling, the secondary air damper modulates to full open to supply air to the space. All the primary air enters the secondary air supply duct.
- When less cooling is required, the secondary air damper to the space modulates towards closed. The excess primary air enters the return air outlet and returns to the central air unit.

CENTRAL AIR SYSTEM

The central air system (Fig. 2) for the bypass system unit is similar to the VAV cooling-only system. Since the system is cooling only, heating is provided by some other means. The primary air is low pressure.

Unoccupied Mode

- ❒ The supply fan and the return fan do not run.
- ❒ The OA (outside air) and EA (exhaust air) dampers are closed. This prevents the coil from freezing at low outside temperatures.
- ❒ The RA (return air) damper is open.

Morning Cool-Down Cycle

- ❒ The cycle starts one to two hours before occupancy.
- ❒ Supply and return fans start.
- ❒ OA, EA, and RA dampers modulate to supply mixed air cold enough to precool the warmest space.
- ❒ The cycle stops when all spaces reach a set temperature.

Occupied Mode

- ❒ Hours of occupied time are preset by clock.
- ❒ Supply and return fans are on continuously.
- ❒ If outside air is above 75°F, the OA damper modulates to the minimum position.
- ❒ If outside air is below 75°F, the mixed air dampers (OA, EA, and RA) provide the primary air duct with the air temperature that will cool the warmest space.
- ❒ If the mixed air temperature is too warm, the control system opens the valve to the chilled water coil to provide the required primary air temperature.

- The primary air discharge temperature will be limited to a minimum temperature (usually around 58°F).

ADVANTAGES OF A BYPASS SYSTEM

- One primary air duct can be used for many zones.
- Installation cost is low.
- If a DX central cooling coil is used, the constant airflow across the coil keeps it from icing up.

DISADVANTAGES OF A BYPASS SYSTEM

- Energy cost is high because of constant primary supply airflow.
- Larger duct and larger fan are required.
- When the damper closes the room supply duct, the air goes to the ceiling opening. There is very little static pressure at the ceiling opening, so the quantity of air from the primary air duct increases. This dumps too much air into the ceiling and may also rob primary air from other terminals.
- If too much pressure builds up in the ceiling, air may flow back into the spaces through ceiling return air outlets. This results in excess cooling in the space.
- A return air fan must be installed in the central air system so that the ceiling space does not become pressurized.

REVIEW

Write your answers to the following questions without referring to the text. Try to answer every question before you check the answers in the back of the book.

1. On the drawing below, add the bypass damper and damper motor with its connections:

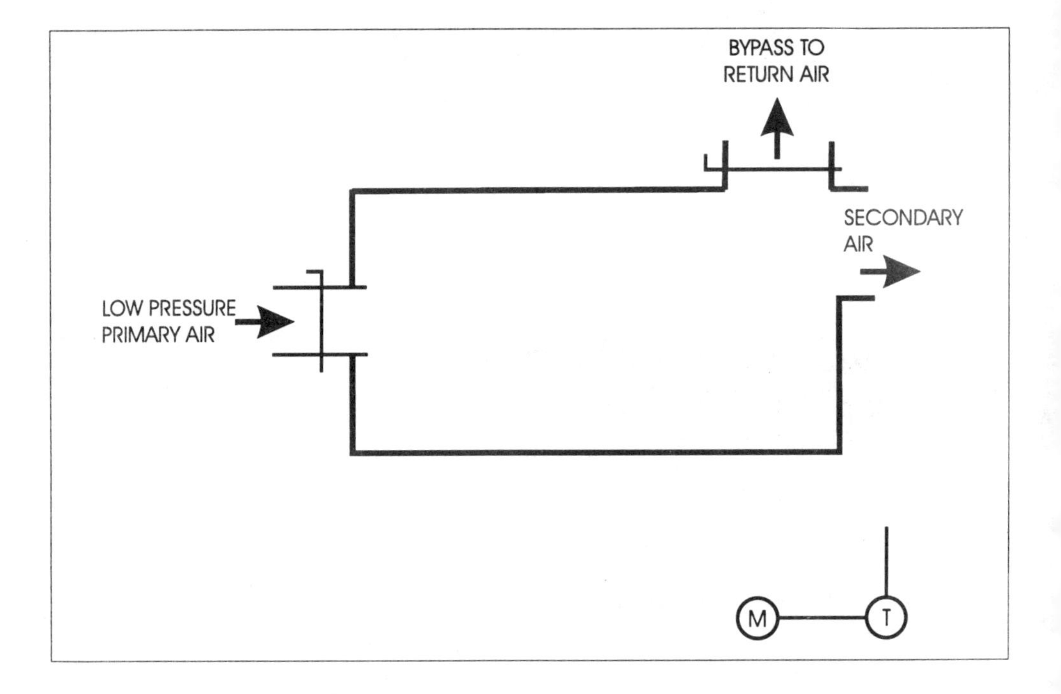

2. Name one advantage of the bypass system.

3. Name one disadvantage of the bypass system.

4. What is the position of the secondary air damper when the space calls for full cooling?

5. Is the primary air fairly high pressure or fairly low pressure?

6. What is the position of the central air unit dampers during the unoccupied mode?
   OA:

   EA:

   RA:

7

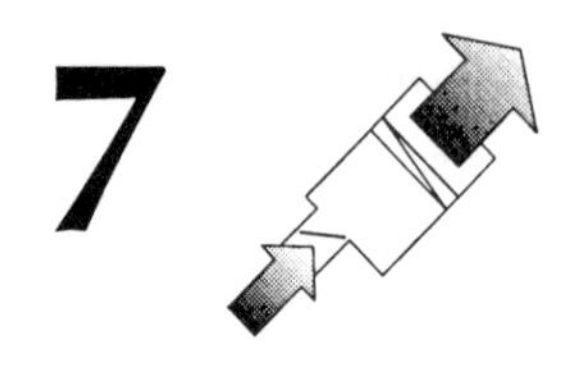

VAV DUAL DUCT SYSTEM

DESCRIPTION

A **VAV dual duct system** has a separate hot duct and cold duct running from the central air unit to each of the terminal units. *This system requires more duct work than other VAV systems.* However, the system is energy-efficient since there is very little bucking, because heating and cooling are not active at the same time. Each terminal unit takes primary air from either the hot duct or the cold duct according to the needs of the space.

COURTESY OF TITUS

VAV DUAL DUCT SYSTEM

Single Duct	Dual Duct	Primary Air		Secondary Air	
		Variable	Constant	Variable	Constant

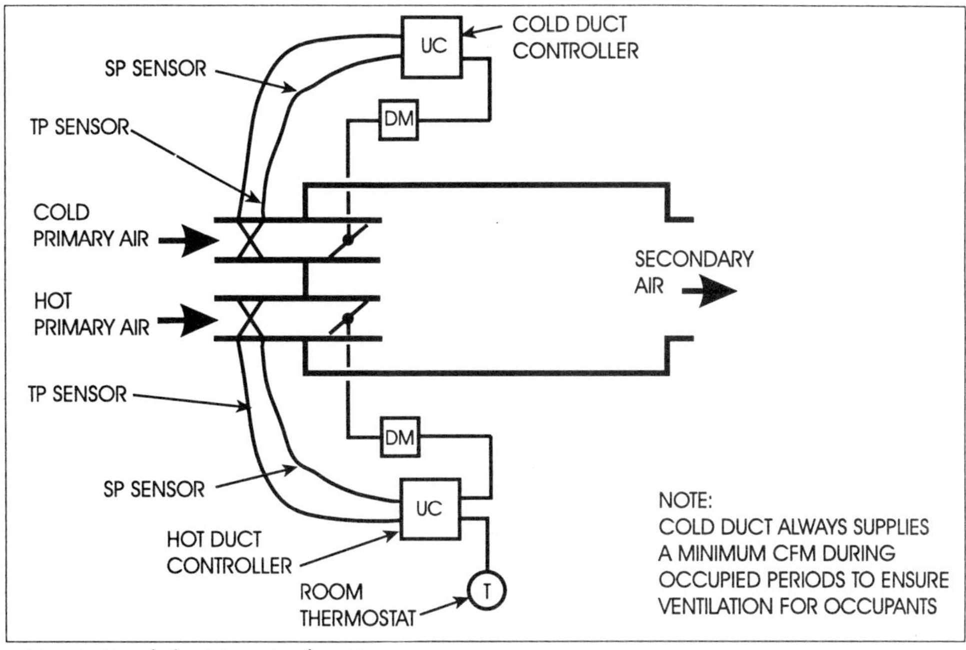

Fig. 1: Dual duct terminal unit

TERMINAL UNIT

Separate primary-air dampers for the hot and cold primary air ducts to the terminal unit operate independently (Fig. 1). These dampers modulate according to a signal from the temperature sensor or thermostat in the space.

The following is a typical sequence of the control system.

Cooling

- ❑ The primary-air damper for the hot primary air duct is closed.
- ❑ The primary-air damper for the cold primary air duct modulates from minimum to maximum to meet the space cooling set point (about 75°F).

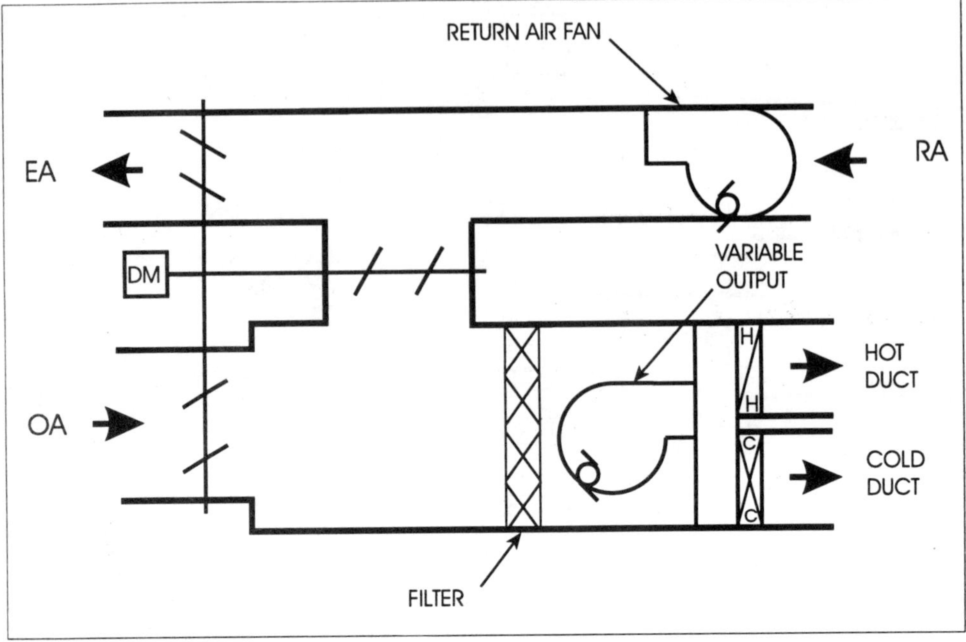

Fig. 2: Dual duct, single-fan central air system

Dead Band

Dead band means that neither heating nor cooling is occurring. Room temperature is between approximately 70°F and 75°F.

- The heated primary-air damper is closed.
- The cold duct primary-air damper is in its minimum position. It must stay partly open to provide the minimum required ventilation air.

Heating

- The hot duct primary-air damper modulates to meet the space heating set point (about 70°F).
- The cold duct primary-air damper is in its minimum position.

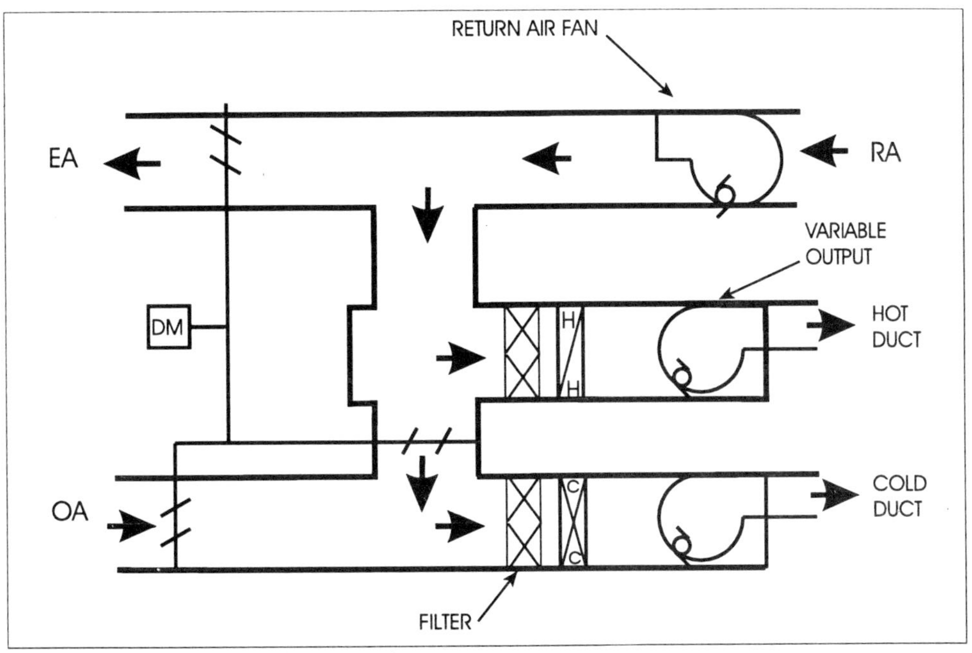

Fig. 3: Dual duct, two-fan central air system

CENTRAL AIR SYSTEM

The central air system for a dual duct system can have either a single supply fan or two fans.

- ❑ The **single-fan unit** (Fig. 2) has a variable speed fan, regulated by an SP sensor in each of the supply ducts to maintain a required minimum static pressure (SP) in the duct with the lowest pressure.
- ❑ The **two-fan unit** (Fig. 3) has separate variable speed fans for the hot duct and the cold duct. These fans operate independently according to the needs of the spaces. The output of each fan is maintained at a proper level to provide the required static pressure in its primary supply duct.

Unoccupied Mode

- ❑ The central supply fans and the return fan do not run.
- ❑ The OA (outside air) and EA (exhaust air) dampers are closed. This prevents the coil from freezing at low outside temperatures.
- ❑ The RA (return air) damper is open.
- ❑ If the temperature in a space falls below a set temperature (usually about 60°F), the supply and return air fans start. The OA and EA dampers remain closed.
- ❑ For a two-fan system, only the heating fan operates.
- ❑ The fans run until the coldest room is satisfied (reaches 60°F).

Morning Warm-up Cycle

- ❑ The cycle starts one to two hours before occupancy.
- ❑ The central supply and return air fans start.
- ❑ For a two-fan system, only the heating fan operates.
- ❑ For a single-fan system, the fan and the heating coil operate.
- ❑ OA and EA dampers remain closed.
- ❑ The cycle stops when all spaces reach the set temperature.

Morning Cool-Down Cycle

- ❑ The cycle starts one to two hours before occupancy.
- ❑ The central supply and return fans start.
- ❑ For a two-fan system, only the cooling fan operates.
- ❑ OA, EA, and RA dampers modulate to supply mixed air cold enough to precool the warmest space.

- Cycle stops when all spaces reach a set temperature.

Occupied Mode

- Hours of occupied time are preset by clock.
- Central supply and return fans are on continuously.
- For a two-fan system, both the cooling fan and the heating fan operate.
- For a single-fan unit, the fan operates, and the cooling and heating coils operate according to demands from the space.
- If outside air is above 75°F, the OA damper modulates to the minimum position.
- If outside air is below 75°F, the mixed air dampers (OA, EA, and RA) modulate to provide the primary air duct with the air temperature that will cool the warmest space.
- If the mixed air temperature is too warm, the control system modulates the valve to the chilled water coil to provide the required primary air temperature.
- The primary air discharge temperature is limited to a minimum temperature (usually around 58°F).
- The RPM of the supply air fan is controlled to maintain the required static pressure (SP) in the supply duct.
- The RPM of the return fan coordinates with the supply fans to maintain a slightly positive building SP (usually around 0.05" wg).

ADVANTAGES OF A DUAL DUCT SYSTEM

- Heating or cooling is available to any space.

- ❑ No piping is required above the ceiling, so there is no danger of leaks.
- ❑ Fans and duct can be smaller.
- ❑ In a two-fan system, the OA goes only to the cooling coil. No OA needs to be heated in the heating cycle.

DISADVANTAGES OF A DUAL DUCT SYSTEM

- ❑ Airflow to the space is variable, and this could affect comfort.
- ❑ Initial cost is higher because there are two duct systems.
- ❑ Controls for the primary air are more complex.

REVIEW

1. What is a disadvantage of the dual duct system?

2. What is the dead band?

3. What is the usual minimum temperature for the primary air?

4. In what mode does the central supply fan not operate?

5. Which damper must stay partly open even when the terminal unit is in the dead band?

6. Does the hot duct or the cold duct have a TP sensor and SP sensor?

8 VAV CHANGEOVER-BYPASS SYSTEM

DESCRIPTION

A VAV **changeover-bypass system** is used to heat and cool small and medium sized buildings. The system provides both heating and cooling to the various zones of the building, but does not do both at the same time. A **bypass damper** maintains a constant static pressure in the primary air duct. The air handler for the system has a constant volume fan, a filter, a heating coil, and a cooling coil in series.

Each zone has a temperature sensor in the space. A master controller communicates with the space temperature sensors as well as with the central air handling unit. It scans the zone sensors on a continuing basis for such information as:

- The amount of deviation from the set temperature
- The length of time the deviation from the set temperature has existed
- The length of time since the last change to the heating or cooling mode
- The number of zones requiring cooling
- The number of zones requiring heating

VAV CHANGEOVER-BYPASS SYSTEM

Single Duct	Dual Duct	Primary Air		Secondary Air	
		Variable	Constant	Variable	Constant

The master controller evaluates the input data and sends a command to the air handler controller to supply either heating or cooling to the zones. If the majority of the spaces require cooling, the air handler goes to the cooling mode. If the majority require heating, the air handler goes to the heating mode.

The controller for the central air handler modulates the outside air, relief air, and return air dampers to provide the temperature of air necessary to best satisfy the commands from the master controller.

CHANGEOVER-BYPASS SYSTEM OPERATION

Figure 1 shows a typical changeover-bypass system.

- ❑ The **fan** is constant volume and runs continuosly when the building is occupied.
- ❑ The **master controller** determines whether the system is to be in the heating mode or cooling mode.
- ❑ The master controller modulates the **bypass damper** for the supply duct to maintain a **constant static pressure** in the primary supply air duct. The bypassed air mixes with the return air from the ceiling plenum and returns to the central air handler. This assures a constant airflow across the cooling coil to prevent it from icing.
- ❑ The outside air and return air dampers modulate to provide the necessary supply air temperature.
- ❑ For each zone, the zone damper modulates to provide the amount of primary air needed to meet heating or cooling requirements.

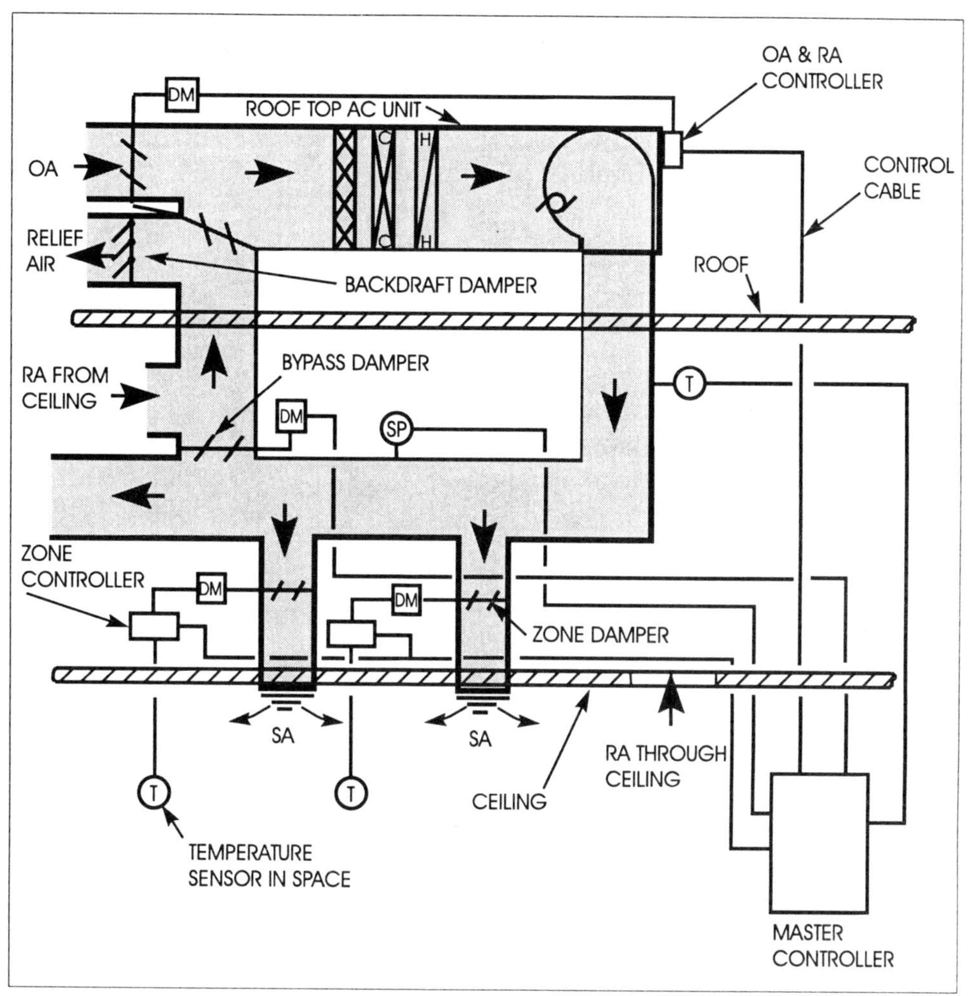

Fig. 1: Changeover-bypass system

- The space above the ceiling is used as a **return air plenum**. The return air is transported back from the plenum to the central air handler.

- A **controller** for each zone modulates a damper in the supply duct to that space. This varies the volume of air being supplied to the zone according to the degree of heating or cooling required. This results in **variable secondary airflow** to the zones.

- The zone controller transmits data to the master controller. The master controller evaluates data from the central air handler, the zone sensors, and the zone controller and sends signals to the zone controller to position the zone damper.

- If the central air handler is in the **heating mode**, and a zone is calling for cooling, the zone controller modulates the zone damper to the minimum position, thus supplying a minimum of air to the zone for ventilation.
- If the central air handler is in the **cooling mode**, and a zone is calling for heating, the zone controller modulates the zone damper to the minimum position. It allows a minimum of ventilation air to the zone.
- The zone controller action reverses according to whether the system is in the heating or cooling mode.

ADVANTAGES OF A CHANGEOVER-BYPASS SYSTEM

- Both heating and cooling are part of the central system, so piping to the terminal units is not needed.
- Zoning is flexible because no terminal units are needed. Terminals are simply duct outlets.

DISADVANTAGES OF A CHANGEOVER-BYPASS SYSTEM

- The system may be constantly changing from heating to cooling.
- Central air handling units should only serve spaces with similar heating and cooling requirements if the system is to work properly. If space needs differ a great deal, more central air handling units may be required.

- ❑ If most of the zones require cooling only, the system may never switch from cooling to heating. The reverse is true if most of the zones require heating only.

REVIEW

1. Where are the heating and cooling coils located?
2. Is the changeover-bypass system more likely to be used for a large building or a small one?
3. Does the central air handler provide heating and cooling at the same time?
4. If one zone for a central air handler requires heating and five require cooling, how does the central air handler respond?
5. Does the fan in the central air handler run constantly?
6. If the central air handler serves five zones, how many dampers are controlled by the master contoller?
7. What is the purpose of the relief air backdraft damper?

9 FAN CONTROL

REASONS FOR FAN CONTROL

Supply air fans for VAV systems need to be controlled so that the fan CFM output matches the system requirements.

In all the systems described in this book, the goal for fan control is to supply the CFM required by the terminal units. This amount can vary from a minimum when all terminal unit dampers are in a minimum position to a maximum when all terminal unit dampers are in a full open position.

An additional goal for fan control is to maintain the lowest possible SP (static pressure) in the primary air duct. With a low SP, the terminal unit dampers do not have to close down as much. The terminal unit dampers should be as open as possible because this reduces pressure loss and air noise.

Early systems had no positive method of fan control. They used a method called **riding the curve**, which means letting the system control itself. As the terminal units modulated their dampers to provide the amount of secondary air needed, they changed the static pressure in the system, and this affected the fan output.

A **fan curve** is a graph provided by the fan manufacturer showing how the fan behaves under specified conditions. For example, it shows what the brake horsepower and static pressure will be at a given output in CFM and a constant RPM. The fan curve also indicates the best operating conditions for the fan. A fan is chosen for a particular application so that the operating conditions will correspond to the best operating conditions on the fan curve. When the fan operating conditions are changed because of the static pressure imposed by the changing terminal unit demands, the operating conditions usually shift to conditions that provide less operating efficiency. This is called **riding the curve**.

The usual result of riding the curve is that increased SP decreases the CFM. This in turn means that the primary-air dampers in the terminal units have to move towards closed. The varying fan output affects the terminal units and causes them to **hunt**, which means to constantly seek the damper position that satisfies the signal from the controller. As a result, the terminal units are difficult to control and balance. This is why some method of fan control is needed.

METHODS OF FAN CONTROL

The CFM (and therefore the SP of the primary air) is commonly controlled in one of these three ways:

- ❑ Using a fan outlet damper
- ❑ Using fan inlet vanes
- ❑ Using a variable frequency drive to control fan speed (RPM))

There are other possible ways to control fan output, but they are seldom used because of the effectiveness of the three commonly used methods. One of these other methods is to change the blade pitch on vaneaxial fans. However, vaneaxial fans are difficult to maintain because of the wear caused by constant modulation of the fan blades.

The three common methods of fan control are covered in this chapter.

Fan Outlet Damper

A **fan outlet damper** was one of the early methods of controlling static pressure (SP). An SP sensor in the primary supply duct sends a signal to a damper motor which modulates a damper at the fan outlet.

This is the least satisfactory of the three methods of fan control. Modulating the damper changes the brake horsepower (Bhp) requirements, the static pressure, and the CFM. Usually the change means less efficient operating conditions for the fan. Using an outlet damper to throttle a fan can cause the fan to become unstable.

Fan Inlet Vanes

Fan inlet vanes (Fig. 1) cause the air to rotate in the direction of the fan wheel. They modulate the amount of air picked up by the fan wheel. Inlet vanes control the CFM entering the fan and therefore control the fan output. Inlet vanes are modulated by a damper motor that receives a signal from an SP sensor in the primary air duct.

This method provides better control than the outlet damper method. However, like the outlet damper method, it also creates less efficient operating conditions for the fan. Inlet vanes do not provide as much control as fan speed control does.

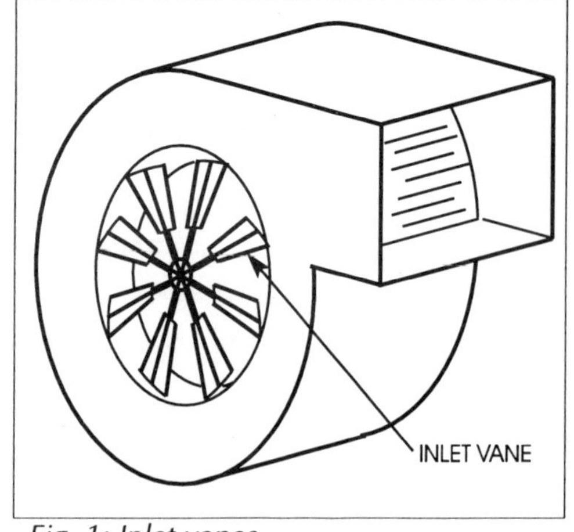

Fig. 1: Inlet vanes

Fan Speed Control with VFD

A **variable frequency drive** (**VFD**) is a popular means for varying fan speed. A **VFD** is an electronic device that changes the frequency (Hertz) of the electrical power to the motor. The VFD varies in size from a small box to a cabinet.

Normal power to a motor is 60 cycle, also called 60 Hertz (Hz). The VFD can vary the frequency of the cycle from 60 Hz down to 10 Hz. This change in frequency changes the motor speed (RPM). The motor used with a VFD must be designed for that use. (If a motor is not designed for use with a VFD, the fan which cools the motor slows down as the motor slows. As a result, the motor overheats.)

The VFD contains a **central processing unit** (**CPU**). This CPU receives the signal from the static pressure sensor in the primary air duct. Acting on this signal, the CPU controls the VFD to change the frequency of the electrical power. This means that the motor adjusts to the speed required to drive the fan so that it will produce the CFM and the SP required for the primary duct.

VFDs automatically improve the power factor. Improved power factor means that the motor draws less current.

The VFD is installed by an electronic specialist. It is located between the motor disconnect switch and the motor. It is an electronic device and should be serviced only by a highly trained specialist. The indoor environment technician will never find it necessary to make any internal adjustments.

If the fan motor is to be stopped, turn off the VFD first before pulling the motor disconnect switch. If the VFD is running when the disconnect switch is pulled, **the VFD can be damaged**.

The use of a VFD can increase **harmonic distortion**. This is simply distortion of the alternating current wave form. Such distortion can affect the operation of electronic devices such as computers. If a VFD affects other equipment in the building, a harmonic filtering device can be installed.

Early VFDs tended to overheat at high elevations. This has been corrected and the VFD is now a highly reliable device.

Depending on the size and model of the VFD, it can have the following features:

- ❑ A soft start, which means that the motor gradually picks up RPM to the desired speed.
- ❑ A digital display that shows the Hertz, volts, and amps to the motor.
- ❑ A manual override to adjust the frequency. This is useful in TAB work.
- ❑ A motor reverse. This is useful in some cooling tower operations.
- ❑ Ability to process either a pneumatic or an electronic input signal to the CPU.

REVIEW

1. For most efficient operation, should the VAV terminal dampers be as open as possible or as closed as possible?

2. For most efficient operation, should the SP in the primary-air duct be as high as possible or as low as possible?

3. Name three common methods of controlling the CFM in the primary-air duct.

4. A VFD (variable frequency drive) can change the frequency of the voltage to a motor to anywhere between 10 and 60 Hz (Hertz). What is the normal frequency for a motor?

5. Does the use of a VFD make the power factor for a building better or worse?

6. What internal adjustments on a VFD are likely to be made by an indoor environment technician?

7. Should a VFD be turned off before or after pulling the disconnect switch for a fan motor?

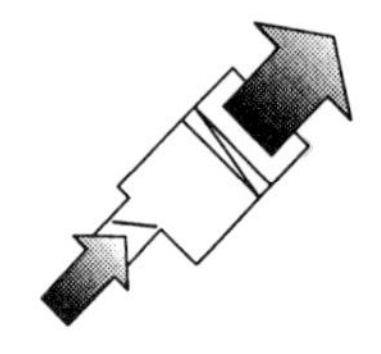

APPENDIX

GLOSSARY

Automatic control system—Control system operated by a central computer

Backdraft damper—Damper that allows air to flow in only one direction

Comfort zone—Range of conditions that satisfy most people who are appropriately dressed and are doing light work, such as office work

Dead band—Range of temperatures where neither heating nor cooling is occurring

Digital control system—Control system operated by a central computer

Digital system—System operated by digital signals (Compare **pneumatic system**)

Economizer cycle—Automatic control cycle that allows the system to use outside air for cooling when this is practical

Fan curve—Manufacturer's graph showing how a fan behaves under specified conditions

Modulate—To vary the degree that a valve or damper is open, somewhere between full open or full closed

Pneumatic control system—Control system operated by signals of varying air pressure (Compare **digital system**)

Primary air—Supply air from the central air handling system

Riding the curve—Letting the HVAC system control the fan operating conditions

Secondary air—Supply air from the VAV terminal unit delivered to a space

Space—The enclosed area that is conditioned by one VAV terminal unit. This is generally one room, but it may be more than one.

Terminal unit—VAV unit

Unit controller—A master control box that receives input signals from sensors and sends signals that control the action of the terminal unit

Variable air volume system—HVAC system designed to meet the needs of different zones in a building

Variable frequency drive—Electronic device that changes fan speed by changing the frequency of the electrical power to the motor

Velocity pressure sensor—A sensor in the duct that sends a signal to the controller to vary the amount of primary air. Also called a **flow cross** or a **velocity pressure probe**

Ventilation—Introducing outside air into a building

Zone—A space or group of spaces in a building with similar requirements for heating and cooling

ABBREVIATIONS USED IN THIS BOOK

Bhp—Brake horsepower

CFM—Cubic feet per minute. Term used for the quantity of air being supplied or exhausted.

CPU—Central processing unit (for computer system)

CV-VT—Constant volume, variable temperature

DM—Damper motor

DX coil—Direct exchange coil that uses refrigerant (rather than water) to cool the air.

EA—Exhaust air

HVAC—Heating, ventilating, air conditioning

HWR—Hot water return

HWS—Hot water supply

OA—outside air

RA—Return air

RPM—Rotations per minute

SP—Static pressure

TP—Total pressure

UC—Unit controller

VAV—Variable air volume

VFD—Variable frequency drive

VV-CT—Variable volume, constant temperature

VV-VT—Variable volume, varying temperature

SYMBOLS USED IN THIS BOOK

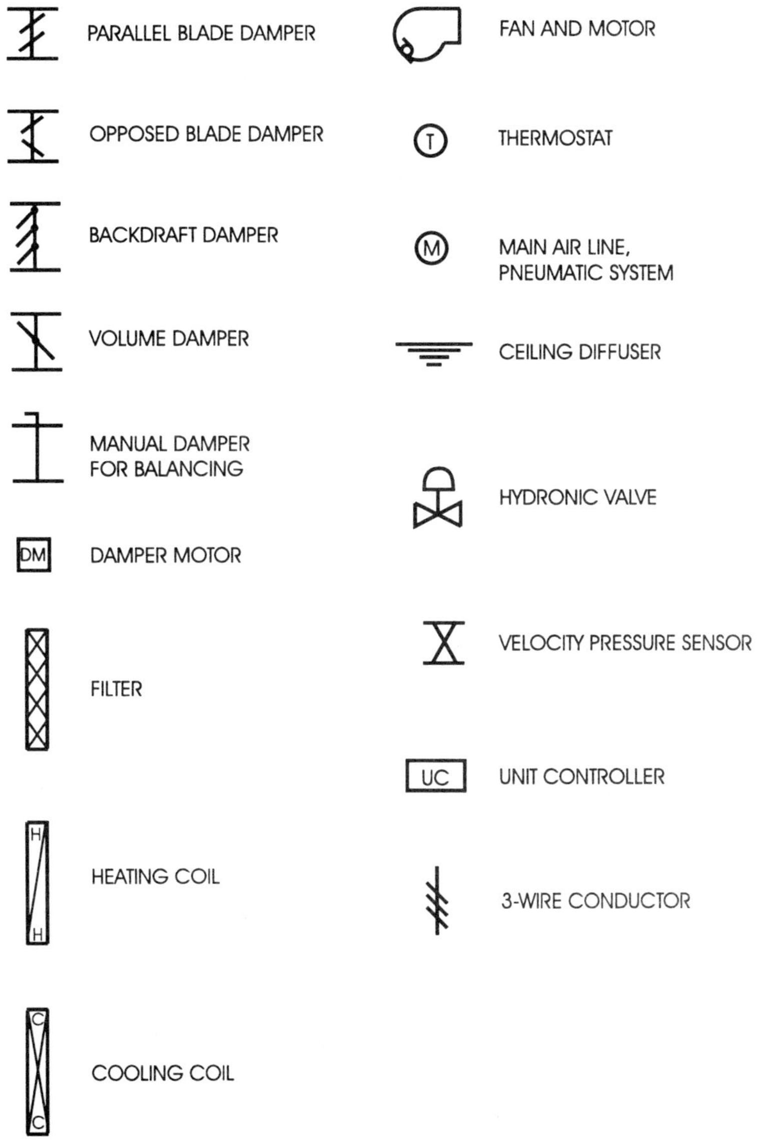

PARALLEL BLADE DAMPER

OPPOSED BLADE DAMPER

BACKDRAFT DAMPER

VOLUME DAMPER

MANUAL DAMPER
FOR BALANCING

DM DAMPER MOTOR

FILTER

H H HEATING COIL

C C COOLING COIL

FAN AND MOTOR

T THERMOSTAT

M MAIN AIR LINE,
PNEUMATIC SYSTEM

CEILING DIFFUSER

HYDRONIC VALVE

VELOCITY PRESSURE SENSOR

UC UNIT CONTROLLER

3-WIRE CONDUCTOR

REVIEW ANSWERS

Chapter 1

1. They rarely need heating.

2. Heat from building transferring to cold outdoor air

3. Heat from outdoor air transferring to building and heat given off by people and equipment in the space

4. Because they use too much energy

5. It saves energy.
   It meets the needs of different spaces.

6. Pressure independent

7. To determine what fan speed is needed

8. Digital

Chapter 2

1. Rooms in the core of the building. Because they are not exposed to cold outside air.

2. Because they gain heat from people and equipment.

3. 55°F

4. Neither. It is the same amount.

5. Slightly positive

6. It measures static pressure and total pressure so that velocity pressure can be determined.

7. One to two hours before occupancy

8. During occupied mode and during morning cool-down

9. OA damper

10. Room temperature and static pressure and total pressure of primary air

11. When the building is occupied and the outside temperature is over 75°F

12. (The drawing should show the RA damper full open and OA and EA dampers closed.)

Chapter 3

1. Static pressure and total pressure from the primary air duct, and room temperature

2. To provide required ventilation air

3. Hot water supply valve

4. Supply and return fans start. OA and EA dampers remain closed. Reheat coils in terminal units are in operation.

5. When the building is occupied

6. During morning warm-up

7. Minimum position

8. Yes

9. (The drawing should show OA and EA dampers full open and RA damper closed.)

10. (See Fig. 1 drawing.)

Chapter 4

1. No
2. Yes
3. To prevent coils from freezing
4. No chance of water leaks
5. Higher energy costs
6. Parallel
7. Series
8. Parallel
9. When calling for heating
10. For both

Chapter 5

1. Constant
2. To draw return air into the terminal unit as needed
3. Just before it enters the terminal unit
4. Minimum

5. Cooling

6. Fairly high

7. Less return air enters the terminal unit

Chapter 6

1. (See Fig. 1 drawing.)

2. Low installation cost

3. High operating cost

4. Full open

5. Fairly low pressure

6. OA: closed
   EA: closed
   RA: open

Chapter 7

1. Higher installation cost because more duct is needed

2. Range at which there is no call for either heating or cooling

3. 60°F

4. Unoccupied mode

5. Cold air damper

6. Both

Chapter 8

1. In the central air handling unit

2. Small building

3. No

4. It provides cooling only to all six zones.

5. Yes

6. 8

7. To prevent pressure from building up in the duct transporting return air and primary air back to the rooftop unit.

Chapter 9

1. As open as possible

2. As low as possible

3. Using a fan outlet damper
   Using fan inlet vanes
   Controlling fan speed

4. 60 Hz

5. Better

6. None

7. Before

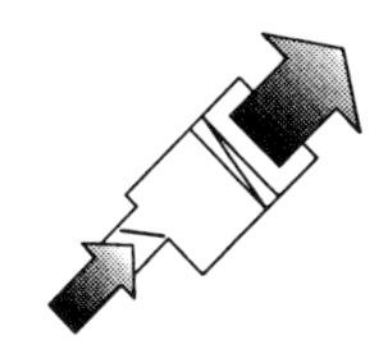

INDEX

LAMA Books also publishes

Teach! *Plain Talk About Teaching*

124 pages, illustrated
7 x 9 softcover - $20.00
ISBN 0-88069-014-3

Teach is an excellent source for anyone teaching.

THIS WILL HELP YOU:

- Teach practical subjects to adults.
- Lead dynamic class discussions
- Manage classroom distractons
- Engross students in learning
- Use media aids effectively
- Write tests and quizzes

The Indoor Environment Technician's Library

Basics of Electricy • Airflow in Ducts • Math for the Technician

100 approx. pages, 6 x 9 softcover - $20.00
Subscribe to the IET series and get the books for $15.00

More than just reference books, the series marks a new way to address the training problems of the industry. Each book focuses on the skills for the technician in heating, ventilating, and air conditioning. May be used in supervised training programs, self-training situations, or just for reference. Books are a handy size, easy reading, with a review at the end of each chapter.

Two different references that cover all the occupational programs in California and the western states.

Occupational Programs in California Public Community Colleges

176 pages 8.5 x 11 softcover
$27.50

Occupational Programs in the Western States

210 pages 8.5 x 11 softcover
$10.00

ORDER FORM

Name ______________________________

Title ______________________________

School/Business ______________________________

Street Address ______________________________

City/State/Zip ______________________________

Telephone ____________ Fax ______________ email ____________

	#	Total
IET **Indoor Environment Technician's Library** (Choose the one you would like to start out with or get any one you wish.) I understand that future titles will be sent periodically on approval and I may return them if they do not fit my purposes. ☐ Please enroll me as a subscriber to the IET Library @ a 20% discount ☐ Please send me more information on the IET Library		
IET titles available: $20.00 non member/$15.00 member's *Basics of Electricity* 7.00 Supervisor's Guide/each		
Airflow in Ducts		
Math for the Technician		
1997-98 Occupational Programs in California Community Colleges $20.00		
Occupational Programs in the Western States $10.00		
TEACH! Plain Talk About Teaching $20.00		
The Reading Program, Books A to G $8.00 each		
Tests and Checkpoints $15.00		
Tapes for Book A $35.00		
Sub total		
California residents add appropriate sales tax		
Add 7% shipping & handling		
TOTAL		

PAYMENT:
- ☐ Check
- ☐ Purchase order
- ☐ Bill me

LAMA Books
Leo A. Meyer Associates Inc.

20956 Corsair Blvd • Hayward CA 94545-1002 • 510-785-1091 • 888-452-6244 • 510-785-1099 Fax • lama@best.com